International Encyclopaedia of Agricultural Engineering

International Encyclopaedia of Agricultural Engineering

Volume 1

R. Maddala

ANMOL PUBLICATIONS PVT. LTD.
NEW DELHI - 110 002 (INDIA)

ANMOL PUBLICATIONS PVT. LTD.

H.O.: 4374/4B, Ansari Road, Darya Ganj,
New Delhi - 110 002 (India)

B.O.: No. 1015, Ist Main Road, BSK IIIrd Stage
IIIrd Phase, IIIrd Block,
Bangalore - 560 085 (India)
Visit us at: www.anmolpublication.com

International Encyclopaedia of Agricultural Engineering

First Published, 2007

ISBN 978-81-261-3375-8 (Set)

PRINTED IN INDIA

Printed at Mehra Offset Press Delhi.

Contents

Preface *vii*

VOL. I

1. Agricultural Engineering: An Introduction 1
2. Agricultural Engineering & Biotechnology: An Introductory Overview 10
3. Agricultural Revolution & Modern Agricultural Practices 104
4. Seed Technology and Plant Growth: Concepts & Approaches 163
5. Agriculture Technology and Food Industry: Tools & Techniques 239
6. Education in Agricultural Engineering, Technology & Communication: Select Case Studies 276

VOL. II

7. Crop Biotechnology & Indian Agriculture: Issues & Options 301
8. Agroecology, Sustainable Farming & Agrobiotechnology: Myths & Risks 346
9. Global Agricultural Agendas & Policy Recommendations 456
10. Agricultural Information Management: Case Study of Select Workshops, Conferences & Publications 522

Glossary 569

Bibliography 577

Index 596

Preface

This publication entitled "International Encyclopaedia of Agricultural Engineering" covers a wide spectrum of related subject matter, such as agriculture engineering, plant biotechnology crop production, plant agriculture, genetic engineering, plant genetic varieties, agro-ecosystems, aquaculture, commercial farming, crop-rotation, horticulture, sustainable agriculture, modern agricultural practices, seed technology, modern food industry, food-safety techniques, agricultural education, agricultural information system, crop-biotechnology, biodiversity, farming systems, global agricultural agenda etc. This wide-spread coverage of issues and themes makes this encyclopaedia a virtual reference book on the said subject. Besides, the encyclopaedia is designed in such a way that it creates understanding of the subject in a systematic manner, so as to greater a much required perspective about this much talked about subject.

The Encyclopaedia provides an introductory overview of agricultural engineering and biotechnology, with focus on the historical evolution of agricultural technologies. Both, agricultural biotechnology and microbiology, have been explained briefly, with focus particularly on fungal biotechnology. The states report on agricultural biotechnology at global level is presented, so are the details of various biotech tools being used to increase crop productivity. Their prospects and limitations in terms of agricultural production applications are covered in detail. Select case studies, such as cotton and maize farming, tobacco growing etc. are explained. Impart assessment of used of biotechnology in agricultural sectors, both as benefits and possible risks, discussed, including the issue of crop-yield sustainability. The relevant situation in India in this regard is also talked about briefly. This publication also presents a picture of the historical development of agricultural revolution and modern agricultural practices. This covers the Neolithic (first agriculture) revolution, green revolution and the role of high-yielding varieties (HYVs), and the Indian, British, Scottish agriculture revolutions. The main principles and practices of modern agricultural system globally are discussed, including processes like horticulture, corporate farming, crop-rotation, short rotation coppice (SRC), alternative crops in double-crop systems, disease resistance in fruits and vegetables, agricultural experiments station, best-agricultural practices and plan for modernization of present agricultural systems. The issue of seed technology, which has attained utmost importance today is also dealt in detail, so are the concepts and approaches of plant reproduction and growth. Special attention is given to various methods for promoting seed germination, with select case studies incorporated herein. Tools and techniques of contemporary agricultural technology and food industry and discussed, including their ecological impact. The focus lies on aquaculture technology, grain cleaners, grain and forage dryers, drying and aeration fans, chemical fertilizer applications, tractor performance monitors, potato planers, etc. An overview of agro-cultural education and extension services is given, with focus on agricultural engineering, technology and communication. Select case-studies are given to further elaborate upon newly developed machinery and automated operating systems in the field of agriculture, including the areas of pest and disease control. All this will help readers understand the academic linkages between various agro-systems, technology and education.

This Encyclopaedia also deals in detail the issue of crop biotechnology and modern agriculture. With particular reference to India. The much talked upon problem of food-safety at global level in also reflected upon briefly. The future of biotech feed crops for livestocks is also explained. The current status of biotechnology and its future challenges,, particularly in India, is dealt briefly. This publication covers extensively the subject areas of agro-ecology, sustainable farming, and role of agro-biotech, focusing mainly on the major benefits, risks and mythos. Select case studies have been mentioned to elaborate the point further. The issue of impact of Climate change and CO2 fertilization on crop yield is also dealt. Ethical questions being raised world-over against use of agricultural biotechnology and the myths surrounding the whole affairs have been addressed. Finally, this encyclopaedia also touches briefly upon the prevailing global agenda towards transforming the agricultural science policy. Relevant chapters of Agenda 21 and details of biodiversity action plan are presented. The Common Agricultural Policy and major policy recommendations for sustainable global agriculture are dealt in detail. An outlook is also given with regard to agricultural information management standards. Select case studies of workshops, conferences and publications are presented, so that readers can gain further insights into this areas of study, with focus on crop, forestry, fishery and animal husbandry sectors. An extensive bibliography, glossary is given, in addition to brief index.

—Editor

1

Agricultural Engineering: An Introduction

Agricultural Engineering

Although agricultural engineering is a fairly new subject, its effect is being felt in India increasingly. In recent years it has contributed to agricultural production in the country both directly and indirectly. The number of modern machines used in agriculture is a fairly good index of the progress made during the four 5 year plans. And yet all this progress is insignificant as compared to that in USSR, USA, Japan, UK Canada, and Australia. This development has made two things clear:

1. That the Indian farmers are quick in adopting new methods and techniques and
2. That the Indian industries can manage as good machines as those abroad.

Agricultural engineers develop engineering science and technology in the context of agricultural production and processing and for the management of natural resources. Agricultural engineers design agricultural machinery and equipment and agricultural structures. Agricultural Engineers may perform tasks as planning, supervising and managing the building of dairy effluent schemes, irrigation, drainage, flood and water control systems, perform environmental impact assessments and interpret research results and implement relevant practices. Some specialties include power system and machinery design; structures and environmental science; and food and bioprocess engineering. They develop ways to conserve soil and water and to improve the processing of agricultural products. A large percentage of agricultural engineers work in academia or for government agencies such as the U.S. Department of Agriculture or state agricultural extension services. Agricultural engineers work in production, sales, management, research and development, or applied science.

Agricultural engineering as defined by international institutions comprises of four main branches, (i) farm implements (ii) rural structures (iii) soil conservation, drainage, and irrigation and (iv) rural electricity.

Because of the peculiar local conditions the lack of electricity on a large scale, and the immediate utility of implements, agricultural engineering in India is more of agricultural implements and machinery. But in fact agricultural engineering is a very vast subject and may be defined as the application of the knowledge, techniques and disciplines of various fields of engineering to the solution of problems arising in the field of agriculture and rural living with the object of reducing labour, improving productivity and raising the standard of living of the farmer and increasing the overall earnings per worker.

The era of more intensive planning brought reorganisation on a wider scale. With the establishment of agricultural universities, several agricultural engineering colleges were established offering graduate and post graduate courses. The most important step has been the establishment of agro-industries corporations with the financial assistance of the Central Ministry of Agriculture. The main objectives of establishing these corporations have been to take over the supply and service functions of the departments of agriculture and to expand them. These agri-industries corporations have each a paid-up capital of 2-5 crores of rupees. They have taken over the existing government workshops also.

The other most important development has been in the field of industry which has so far half a dozen factories for manufacturing tractors, 3 or 4 for power tillers, and several for oil engines and electrically driven pumpimg sets. Some of these are being assisted by the Industrial Finance Corporation of India. These helped to quicken the pace of mechanisation of Indian agriculture.

It is not merely the knowledge of agricultural engineering which is needed for increasing production, but also the knowledge of their availability, prices and upkeep. Information can be obtained from:

1. At the village level: the village level worker or gram sevak.
2. At the block level: the Agricultural Extension Officer or the District Planning Officer.
3. At the district level: The District Agricultural Officer or the District Planning Officer.
4. At the state level: the State Agricultural Engineering Section and the Managing Director of the State Agro-Industries Corporation.
5. At the Central level: the Director of Agricultural implements and Joint Commissioner (machinery), Ministry of Agriculture and Irrigation, Krishi Bhavan, New Delhi.

The private industry as well as the government, through their agro-industries corporations are in the field now for the supply of agricultural machinery to farmers. To prevent the charging of exorbitant prices for tractors, the govt has fixed the price of tractors with the assistance of the Bureau of Cost Accounting in the Ministry of Finance. This organisation has helped the farmers very much.

Agricultural aviation is now gaining ground in India. At present the aeroplanes and helicopters are being used mainly for spraying pesticides but in future they may be used for sowing and spraying liquid fertilizers on the crops. Agricultural electronics has also begun and remote-control tractors are being designed and tested in some countries. With the successful explosion of an atomic device India may be able to generate more electricity in the future. The availability of plentiful irrigation water as a result of the Ganga-Kauveri project and the availability of more electricity from atomic energy will make the progress of farm mechanisation in India speedier. Thus the future seems to be bright.

Agricultural engineers apply knowledge of engineering technology and science to agriculture and the efficient use of biological resources. They design agricultural machinery and equipment and agricultural structures. Some specialize in areas such as power systems and machinery design; structures and environment engineering; and food and bioprocess engineering. They develop ways to conserve soil and water and to improve the processing of agricultural products. Agricultural engineers often work in research and development, production, sales, or management.

Agricultural Implements and Machinery

The beginning of mechanisation of Indian agriculture was made by the use of improved hand tools and improved bullock-drawn implements. Particular stress was laid on these items during the first 3 Five year plans and slowly more sophisticated implements were introduced. These included power-driven threshers, tractors and the combine harvestors. Quite a variety of implements is required for carrying agricultural operations, starting from the opening of the land to harvesting, threshing, winnowing and storage. These operations include ploughing, harrowing, levelling, sowing, intercultivation, application of fertilizers and manures, harvesting, threshing and winnowing. A variety of *desi* implements are in use in the country for the last so many centuries. Before efforts were made to introduce foreign implements a survey implements used in India was taken. Therefore under the Indian Council of Agricultural Research a country wide survey was carried out and its report was published. This has brought many facts to light. The number of implements used by the Indian farmer is very large. Some of these implements are crude to look at but are very useful and the consensus of the agricultural engineers of India has been that it would be easier at least as a first step to improve upon the indigenous implements in order to increase their efficiency and thus give to the Indian farmers something which is within their means. To achieve this objective effectively, the Indian Council of Agricultural research has now established one research-cum testing and training centre in each state. Some of them have now been converted into regional centres or have been taken over by the state governments.

The object is to improve the indigenous agricultural implements, to design new implements and to test them in the field upto the prototype level. After the implements have proved to be useful, the prototypes can be given to manufacture in public or private sectors and the implements can be sold through them to the farmers. Principles of research. In India, the holdings are generally small, the average farmer is poor, and the draft cattle are also small and ill fed. Taking these into account certain principles have been laid down which are:

1. That the implements and machinery for the Indian farmers should be simple in construction, so that they could be operated by the illiterate farmers, and should either be manufactured are could be repaired by the village artisans or mechanics.
2. The prices of the implements should be within reach of the ordinary farmer. In case of power-driven equipment the difficulty of high cost can be overcome by giving the machinery on hire or hire-purchase system or on custom basis or on cooperative basis.
3. The implements should be light so that they can be easily transported and are suitable for the draft animals.
4. That the implements should be scientifically tested on a fairly uniform basis throughout India.
5. In designing the implements the local available materials are to be used. In the case of soil working and plant working parts durable materials such as carbon steel should be used.

To fulfil these requirements, the Indian Council of Agricultural Research has also designed procedures for testing and designing implements and preforms in which the test data are to be listed. Quite a large number of farmers are using the power machines like tractors, oil engines and pumps. Thus the trend towards mechanised farming is bound to increase.

Recently, new developments have taken place to bring about mechanization of agriculture. Owing to the shortage of diesel oil and its increased price, the mechanical operation will have to be restricted to deep ploughing, land levelling, clearing land and other operations which cannot be carried out by bullock power. For short distances bullock carts are more effective than trucks; hence the necessity for improving them. Among the indigenous implements many seed sowing devices are attached to the implements such as the desi plough. They are commonly called nari plough, tifan, argada and other multi-type elements that are used in the Deccan. Tubes are fixed to these implements and on the top of these a seedling bowl into which the seed is dropped is fixed. Thc seedling bowl has a number of holes in it. The construction of the holes is ingenuous with the result that the seed falls right into the holes and into the tines. In some agriculturally backward areas, broadcasting or scattering the seeds randomly is also followed but the major portion of the land is now sown in lines and the seed is drilled. Sowing in lines ensures the proper distribution and placing of seeds, resulting in the uniform spacing of the plant with a concomitant reduction in seed rate. It also helps the seed to germinate regularly, ensuring a better standard of the crop and also facilitates interculture with bullock-operated holes thus reducing the cost of cultivation. Quite a large number of improved seed drills automatic and hand-operated have been designed. Among the hand designed ones, a copy of the wooden tifan has been made in Uttar Pradesh. It is made of mild steel with an arrangement for changing the distance between the tines. Among the automatic seed-drills, the Mc Cormic seed-drill, the Master seed drill, the Gunti seed drill and the Kirloskar seed drill may be mentioned. Most of the manufacturers now try to combine the simultaneous feeding of fertilizer with the dropping of the seed, thus making the implement a seed cum fertilizer drill. So far, none of these combined devices has proved very popular because of high cost, complicated design, greater draft etc.

The seed drills need a lot of skill in operation. It is necessary in case of automatic seed drills to caliberate them before sowing to know the exact seed rate it will drop. The seed drills are to be caliberated for each type of crop separately because of differences in shape, size and weight of different seeds. Proper care has to be taken in choosing the soil working part such as the tine. If the soil is hard, strong straight tines with sharp edges should be selected. If the soil is soft, the disc shovels may be preferred. Some other factors to be kept in mind are the covering of the seeds, the pressing of soil slightly over the seeds, the transport facilities for the drills etc.

Among the sowing devices mention can also be made of the transplanting machines. Even though streneous efforts are being made in Japan, Italy, USA and China, so far no really useful machine for transplanting paddy has been made except in Japan. Therefore transplanting has to be done by hand. A small tool for transplanting 2 or 3 seedlings of paddy at a time has been developed in Burma and Malaysia. There is a need to introduce this tool in India because with a little practice it can increase the efficiency by 20%.

Most of the inter culture in India is done with the hand tool called khurpi which consists of a sharp edged triangular or sickle-shaped blade. Sometimes weeding is done with a small spade, hoe or mummuty. Most of these tools have small handles and the user has to sit on his heels or bend his back while weeding.

Weeding is one single operation which can increase crop yields without much additional investment. To increase the efficiency of weeding, long hand lead hoes such as the Singh Hand hoe, the Sharma Hand hoe and the Dutch Hand hoe have been introduced. The price is Rs.6 and is well within means even for the poor farmer. They can weed 4 or 5 times more than a khurpi. These hand hoes are simple in construction and can be made by the village blacksmith. They are thus strongly recommended.

The wheel hoes are a further improvement upon the hand hoes. A wheel hoe consists of a body, handle and a wheel. To its body many tools like a shovel, a ridger, and other types of tools can be fixed. The price is Rs 60 to 75 with a weeding attachment. Other attachments can be purchased at no extra cost. Being provided with a wheel it is easier to work and very useful for intercultivation in vegetable gardens and orchards. Where the soil is soft it requires only one man to work at this.

For hoeing line sown crops with bullock power a light desi plough is used in northern India. In the area between the Narmada and the Krishna river, multi-tined bullock operated hoes are used. They are known as doura, douri, dundia and dulari. These are the miniature firms of bakhars. When the plants are small multi tined hoes fixed to a single wooden body are also used in Andhra Pradesh and Karnataka. The draft of these hoes is very light and sometimes 4 of them can be pulled by a single bullock. The blades work on each side of a row of plants. In the case of rain fed crops hoeing by using bullocks is done at most 3-4 times after which the plants grow tall and bushy and the soil is also hard. To cope with this stronger and harder blades have been designed. They are commonly known as the Washim hoe, the Akola hoe, the Baroda hoe and the triphali. Their designs are similar and are very helpful for the interculture of cotton crop grown in rows.

Yet another implement called the Indore ridger was introduced in Malwa some years back. This has a sharp beam with sloping ends and a hexagonal share attached to it with a strong iron bar. The ridger can be used up to a later stage of the crop without doing physical injury to the plants. It not only eradicates weeds but also produces a good mulch. Similar to the doura 2 or 3 such ridgers can be attached to a pair of bullocks. There is scope to introduce these implements in northern India where they have not yet been introduced so far.

For some special crops like paddy, special types of paddy weeders have been designed. They are based on the Japanese design. They consist of 2 rollers with pointed fingers and are pushed in between the rows of the paddy plants when water is standing in the fields. The idea is to bury the weeds in the mud to decompose them and add organic matter to the soil.

With the advancement of scientific research new methods to control weeds with selective regicides are being used. When this regicide is sprayed in between the crops it kills the weeds only and not the crops. At present they are costly but when they become abundant, they will be more effective than other implements. Some of the intercultural implements may have some attachments fixed to them to carry out some special jobs such as sulphur dusting in tobacco cultivation and the top-dressing of fertilizers. Already simple sulphur dusters which can be worked in rows and fertilizer attachments to the weeders have been designed..

Agricultural Science

Agricultural science is a broad multidisciplinary field that encompasses the parts of exact, natural, economic, and social sciences that are used in the practice and understanding of agriculture. (Veterinary science, but not animal science, is often excluded from the definition.)

Agriculture and Agricultural Science

The two terms are often confused. However, they cover different concepts:

- Agriculture is the set of activities that transform the environment for the production of animals and plants for human use. Agriculture concerns techniques, including the application of agronomic research.
- Agronomy is research and development related to studying and improving plant-based agriculture.

Agricultural sciences include research and development on:

- Production techniques (e.g., irrigation management, recommended nitrogen inputs)
- Improving agricultural productivity in terms of quantity and quality (e.g., selection of drought-resistant crops and animals, development of new pesticides, yield-sensing technologies, simulation models of crop growth, in-vitro cell culture techniques)
- Transformation of primary products into end-consumer products (e.g., production, preservation, and packaging of dairy products)
- Prevention and correction of adverse environmental effects (e.g., soil degradation, waste management, bioremediation)
- Theoretical production ecology, relating to crop production modeling
- traditional agricultural systems such as which serve to feed most people in the world and which often retain integration with nature in a way that has proven more sustainable than modern systems
- Food production and demand on a global basis, with special attention paid to the major producers of China and India.

Agricultural Science: A Local Science

With the exception of theoretical agronomy, research in agronomy, more than in any other field, is strongly related to local areas. It can be considered a science of eco-regions, because it is closely linked to soil properties and climate, which are never exactly the same from one place to another. Many people think an agricultural production system relying on local weather, soil characteristics, and specific crops has to be studied locally. Others feel a need to know and understand production systems in as many areas as possible, and the human dimension of integration with nature.

History of Agricultural Science

Agricultural science is seen by some to have began with Mendel's insightful genetic work, but in modern terms might be better dated from the chemical fertilizer outputs of plant physiological understanding in eighteenth century Germany. Today it is very different from what it was even in 1950. Intensification of agriculture since the 1960s in developed and developing countries, often referred to as the Green Revolution, was closely tied to progress made in selecting and improving crops and animals for high productivity, as well as to developing additional inputs such as artificial fertilizers and phytosanitary products.

As the oldest and largest human intervention in nature, the environmental impact of agriculture in general and more recently intensive agriculture, industrial development, and population growth have raised many questions among agricultural scientists and have led to the development and emergence of new fields. These include technological fields that assume the solution to technological problems lies in better technology, such as integrated pest management, waste treatment technologies, landscape architecture, genomics, and agricultural philosophy fields that include references to food production as something essentially different from non-essential economic 'goods'. In fact, the interaction between these two approaches provide a fertile field for deeper understanding in agricultural science.

New technologies, such as biotechnology and computer science (for data processing and storage), and technological advances have made it possible to develop new research fields, including genetic engineering, agrophysics, improved statistical analysis, and precision farming. Balancing these, as above, are the natural and human sciences of agricultural science that seek to understand the human-nature interactions of traditional agriculture, including interaction of religion and agriculture, and the non-material components of agricultural production systems.

Prominent Agricultural Scientists

- Norman Borlaug
- Luther Burbank
- Louis Pasteur

- Gregor Mendel
- René Dumont
- George Washington Carver

Agricultural Science and Agriculture Crisis

Agriculture sciences seek to feed the world's population while preventing biosafety problems that may affect human health and the environment. This requires promoting good management of natural resources and respect for the environment, and increasingly concern for the psychological well-being of all concerned in the food production and consumption system.

Economic, environmental, and social aspects of agriculture sciences are subjects of ongoing debate. Recent crises (such as Avian Flu, mad cow disease and issues such as the use of genetically modified organisms) illustrate the complexity and importance of this debate.

Fields or Related Disciplines

- Agricultural engineering
- Agricultural philosophy
- Biosystems engineering
- Aquaculture
- Agronomy and Horticulture
- Agrophysics
- Animal science
- Plant fertilization, animal and human nutrition
- Plant protection and animal health
- Soil science
- Water science
- Biotechnology, genetic engineering, and microbiology
- Farming equipment
- Irrigation and water management
- Agricultural economics
- Food science
- Environmental science and engineering
- Waste management
- Ecology and environment
- Theoretical production ecology

2

Agricultural Engineering & Biotechnology: An Introductory Overview

Agricultural biotechnology has been advancing rapidly; and for all the promises it offers, it poses as many questions. Agricultural biotechnology is rewriting the rules in several key areas—agricultural research policy, industry structure, production and marketing, consumer preference, and world food demand—and public policy is struggling to keep up. Much of the current interest in biotechnology stems from the rapid diffusion in North America and other exporting countries, like Argentina, of genetically engineered (GE) crops such as cotton, soybeans, corn, and canola, and from the uneasy consumer response in Europe as compared with the United States.

"Broadly speaking, biotechnology is any technique that uses living organisms or substances from these organisms to make or modify a product for a practical purpose. Biotechnology can be applied to all classes of organism — from viruses and bacteria to plants and animals — and it is becoming a major feature of modern medicine, agriculture and industry. Modern agricultural biotechnology includes a range of tools that scientists employ to understand and manipulate the genetic make-up of organisms for use in the production or processing of agricultural products.

Some applications of biotechnology, such as fermentation and brewing, have been used for millennia. Other applications are newer but also well established. For example, micro-organisms have been used for decades as living factories for the production of life-saving antibiotics including penicillin, from the fungus Penicillium, and streptomycin from the bacterium Streptomyces. Modern detergents rely on enzymes produced via biotechnology, hard cheese production largely relies on rennet produced by biotech yeast and human insulin for diabetics is now produced using biotechnology.

Biotechnology is being used to address problems in all areas of agricultural production and processing. This includes plant breeding to raise and stabilize yields; to improve resistance to pests, diseases and abiotic stresses such as drought and cold; and to enhance the nutritional content of foods. Biotechnology is being used to develop low-cost disease-free planting materials for crops such as cassava, banana and potato and is creating new tools for the diagnosis and treatment of plant and animal diseases and for the measurement and conservation of genetic resources. Biotechnology is being used to speed up breeding programmes for plants, livestock and fish and to extend the range of traits that can be addressed. Animal feeds and feeding practices are being changed by biotechnology to improve animal nutrition and to reduce environmental waste. Biotechnology is used in disease diagnostics and for the production of vaccines against animal diseases.

Clearly, biotechnology is more than genetic engineering. Indeed, some of the least controversial aspects of agricultural biotechnology are potentially the most powerful and the most beneficial for the poor. Genomics, for example, is revolutionizing our understanding of the ways genes, cells, organisms and ecosystems function and is opening new horizons for marker-assisted breeding and genetic resource management. At the same time, genetic engineering is a very powerful tool whose role should be carefully evaluated. It is important to understand how biotechnology — particularly genetic engineering — complements and extends other approaches if sensible decisions are to be made about its use.

This chapter provides a brief description of current and emerging uses of biotechnology in crops, livestock, fisheries and forestry with a view to understanding the technologies themselves and the ways they complement and extend other approaches. It should be emphasized that the tools of biotechnology are just that: tools, not ends in themselves. As with any tool, they must be assessed within the context in which they are being used."

Biotechnology *is the application of scientific techniques to modify and improve plants, animals, and microorganisms to enhance their value. Agricultural biotechnology is the area of biotechnology involving applications to agriculture.* Agricultural biotechnology *has been practiced for a long time, as people have sought to improve agriculturally important organisms by selection and breeding. An example of traditional agricultural biotechnology is the development of disease-resistant wheat varieties by cross-breeding different wheat types until the desired disease resistance was present in a resulting new variety.*

In the 1970s, advances in the field of molecular biology provided scientists with the ability to readily transfer DNA — the chemical building blocks that specify the characteristics of living organisms — between more distantly related organisms. Today, this technology has reached a stage where scientists can take one or more specific genes from nearly any organism, including plants, animals, bacteria, or viruses, and introduce those genes into another organism. This technology is sometimes called *genetic engineering*. An organism that has been modified, or transformed, using modern biotechnology techniques of genetic exchange is referred to as a genetically modified organism ("GMO"). Actually genetic modification has been around for hundreds if not thousands of years, since deliberate crosses of one variety or breed with another result in offspring that are genetically modified compared to the parents, and hybrid crosses result in progeny with genetic combinations of closely related species.

What is Agricultural Biotechnology?

The Convention on Biological Diversity (CBD) defines biotechnology as: "any technological application that uses biological systems, living organisms, or derivatives thereof, to make or modify products for specific use" (Secretariat of the Convention on Biological Diversity, 1992). This definition includes medical and industrial applications as well as many of the tools and techniques that are commonplace in agriculture and food production.

The Cartagena Protocol on Biosafety defines "modern biotechnology" more narrowly as the application of:

1. In vitro nucleic acid techniques, including recombinant deoxyribonucleic acid (DNA) and direct injection of nucleic acid into cells or organelles, or
2. Fusion of cells beyond the taxonomic family, that overcome natural physiological reproductive or recombination barriers and that are not techniques used in traditional breeding and selection.

The FAO Glossary of biotechnology defines biotechnology broadly as in the CBD and narrowly as "a range of different molecular technologies such as gene manipulation and gene transfer, DNA typing and cloning of plants and animals" (FAO, 2001a).

Recombinant DNA techniques, also known as genetic engineering or (more familiarly but less accurately) genetic modification, refer to the modification of an organism's genetic make-up using transgenesis, in which DNA from one organism or cell (the transgene) is transferred to another without sexual reproduction. Genetically modified organisms (GMOs) are modified by the application of transgenesis or recombinant DNA technology, in which a transgene is incorporated into the host genome or a gene in the host is modified to change its level of expression. The terms "GMO", "transgenic organism" and "genetically engineered organism (GEO)" are often used interchangeably although they are not technically identical. For the purposes of this report they are used as synonyms."

Understanding, Characterizing and Managing Genetic Resources

The Source Document used for this study states: "Farmers and pastoralists have manipulated the genetic make-up of plants and animals since agriculture began more than 10 000 years ago. Farmers managed the process of domestication over millennia, through many cycles of selection of the best adapted individuals. This exploitation of the natural variation in biological organisms has given us the crops, plantation trees, farm animals and farmed fish of today, which often differ radically from their early ancestors).

The aim of modern breeders is the same as that of early farmers — to produce superior crops or animals. Conventional breeding, relying on the application of classic genetic principles based on the phenotype or physical characteristics of the organism concerned, has been very successful in introducing desirable traits into crop cultivars or livestock breeds from domesticated or wild relatives or mutants (Box 3). In a conventional cross, whereby each parent donates half the genetic make-up of the progeny, undesirable traits may be passed on along with the desirable ones, and these undesirable traits may then have to be eliminated through successive generations of breeding. With each generation, the progeny must be tested for its growth characteristics as well as its nutritional and processing traits. Many generations may be required before the desired combination of traits is found, and time lags may be very long, especially for perennial crops such as trees and some species of livestock. Such phenotype-based selection is thus a slow, demanding process and is expensive in terms of both time and money. Biotechnology can make the application of conventional breeding methods more efficient."

Induced Mutation-assisted Breeding

"Spontaneous mutations are the "natural" motor of evolution, and the resource into which breeders tap to domesticate crops and to "create" better varieties. Without mutations, there would be no rice, or maize or any other crop.

Starting in the 1970s, the International Atomic Energy Agency (IAEA) and FAO sponsored research on mutation induction to enhance genetic improvement of food and industrial crops for breeding new improved varieties. Induced mutations are brought about by treating plant parts with chemical or physical mutagens and then selecting for desirable changes — in effect, to mimic spontaneous mutations and artificially broaden genetic diversity. The precise nature of the mutations induced has generally not been a concern irrespective of whether the mutant lines were used directly or as sources of new variation in cross-breeding programmes.

Induced mutation to assist breeding has resulted in the introduction of new varieties of many crops such as rice, wheat, barley, apples, citrus, sugar cane and banana (the FAO/IAEA Mutant Varieties Database lists more than 2 300 officially released varieties). The application of mutation induction to crop breeding has translated into a tremendous economic impact on agriculture and food production that is currently valued in billions of US dollars and millions of hectares of cultivated land. Recently, mutation techniques have undergone a renaissance, expanding beyond their direct use in breeding into novel applications such as gene discovery and reverse genetics.

Features of Agricultural Biotechnology

The First Decade of Genetically Engineered Crops in the United States—Over the past ten years, farmers have adopted genetically engineered (GE) varieties of corn, soybeans, and cotton widely and at rapid rate and benefited from such adoption. While the level of consumer concerns about foods that contain GE ingredients varies by country, with European consumers being most concerned, these concerns have not had a large impact on the market for these foods in the United States.

Use of Crop Genetically Engineered Crops Rising—Driven by farmers' expectations of higher yields, savings in management time, and lower pesticide costs, the adoption of first-generation genetically engineered crop varieties with enhanced input traits has increased rapidly. Adoption of genetically engineered soybeans, corn, and cotton by U.S. farmers has climbed most years since introduction. See the Amber Waves Summary.

Ag Biotech Patents on the Move—Beginning in the late 1990s, the larger agricultural biotechnology companies began acquiring the smaller ones. Mergers among several of the large firms placed a majority of agbiotech patents in the hands of a dwindling number of large, international corporations. This concentration of patent ownership means that an inc easing share of future research will probably be done by companies with the large scale necessary to handle technology development, product marketing, and regulation compliance efficiently. But these companies might restrict research to complement their existing products. (6/05)

Genetically Engineered Crop Varieties Gain Further Acreage Share in 2004—Genetically engineered varieties of soybeans, corn, and cotton have been available commercially since 1996. Since then, their rate of use by U.S. farmers has climbed most years, including 2004. For the most part, farmers have adopted herbicide-tolerant (HT) varieties, which help control weeds, at a faster pace than insect-resistant varieties, which help control insects. (09/04)

Have Seed Industry Changes Affected Research Effort?—Crop variety R&D over the past 30 years has moved from being predominantly public to predominantly private. Private sector firms have evolved from small operations to large integrated enterprises capable of variety development and seed production, conditioning, and marketing. Greater protection of intellectual property rights for crop-seed innovations through patents and certificates has spurred private investment in general and may increasingly stimulate private R&D. Still, ERS analysis shows that consolidation in the private seed industry over the past decade may have dampened the intensity of private research undertaken on crop biotechnology relative to what would have occurred without consolidation, at least for corn, cotton, and soybeans. (02/04)

The Seed Industry in U.S. Agriculture—Unprecedented growth in crop yields and agricultural total factor productivity over the past 70 years owes much to biological innovation embodied in seeds, beginning with the development of hybrid crops in the United States in the early part of the 20th century, continuing with the Green Revolution of the 1960s and early 1970s, and, more recently, modern biotechnology. Throughout this period, the seed industry has evolved. This publication explores data and information on crop seed markets, regulation, industry, and R&D relating to the U.S. seed industry.

Consumers and the Future of Biotech Foods in the United States—Biotechnology is often associated with promise to feed the world, promise to reduce environmental harm, promise to expand agricultural markets and production possibilities, promise to create products that consumers want. In light of these potential applications, this article examines the level of acceptance of biotechnology by farmers and consumers in the U.S.

The Effects of Information on Consumer Demand for Biotech Foods: Evidence from Experimental Auctions—Consumers' willingness to pay for food products decreases when the food label indicates that a food product is produced with the aid of modern biotechnology. An experimental auction was designed and conducted to elicit consumers' willingness to pay for "genetically modified" (GM)-labeled and standard-labeled foods under different information regimes. For vegetable oil, tortilla chips, and potatoes, labels mattered. Consumers discounted food items labeled "GM" by an average of 14 percent. The study also found that gender, income, and other demographic characteristics had only a slight impact on consumers' willingness to pay for biotech foods, while information from interested parties and independent third-party sources had a strong impact. (4/03)

Adoption of Bioengineered Crops—This report uses USDA survey data to examine the extent to which US farmers have adopted bioengineered crops, factors affecting adoption of these crops, and the impacts of bioengineered crops on input use and farm-level net returns. (6/02)

Size and Distribution of Market Benefits from Adopting Biotech Crops—This report estimates the size and distribution of market benefits from adopting Bacillus thuringiensis (Bt) cotton, herbicide-tolerant cotton, and herbicide-tolerant soybeans in 1997. Impacts on U.S. farmers, U.S. consumers, biotechnology developers, germplasm suppliers, and producers and consumers in other countries are assessed.

Economic Issues in Agricultural Biotechnology—The emergence of agricultural biotechnology introduces new concerns about the supply of new technology and its impact on the marketplace New developments in marketing and contractual arrangements between farmers and grain and food processors will likely become critical issues as more genetically engineered (GE) crops enter the marketplace. This report explores some implications for grades and standards for GE products, as well as the extent of biotechnology adoption by U.S. farmers and some of the farm-level effects. Other topics include important advances in biological science, the roles of public and private research, and recent changes in input industry structure. Also examined are consumer preferences, particularly U.S. versus European, and biotechnology's potential to feed a growing world population. (3/01)

Related Data Products

Agricultural Biotechnology Intellectual Property—This database identifies and describes U.S. utility patents on inventions in biotechnology and other biological processes-with issue dates between 1976 and 2000—that are used in food and agriculture. The database also provides information about the ownership of these patents, whether patents are held in the public or private sector, and changes in patent ownership due to firm mergers, acquisitions, and spinoffs.

Genetically Engineered Crops in the U.S.—This data product provides information on the extent of adoption of genetically engineered herbicide-tolerant and insect-resistant crops in the United States 2000-2006.

Microbiology

Microbiology is the study of *microorganisms*, which are unicellular or cell-cluster microscopic organisms. This includes eukaryotes such as fungi and protists, and prokaryotes such as bacteria and certain algaes. Viruses, though not strictly classed as living organisms, are also studied. People that study the field of microbiology are known as *microbiologists*.

Although much is now known in the field of microbiology, advances are being made regularly. The most common estimates suggest that we have studied only about 1% of all of the microbes in any given environment. Thus, despite the fact that over three hundred years have passed since the discovery of microbes, the field of microbiology is clearly in its infancy relative to other biological disciplines such as zoology, botany and entomology.

Bacteria were first observed by Anton van Leeuwenhoek in 1676 using a single-lens microscope of his own design. The name "bacterium" was introduced much later, by Ehrenberg in 1828, derived from the Greek meaning "small stick". While Antony van Leeuwenhoek is often cited as the first microbiologist, the first recorded microbiological observation, that of the fruiting bodies of molds, was made earlier in 1665 by Robert Hooke.

The field of bacteriology (later a subdiscipline of microbiology) is generally considered to have been founded by Ferdinand Cohn (1828-1898), a botanist whose studies on algae and photosynthetic bacteria led him to describe several bacteria including

Bacillus and *Beggiatoa*. Ferdinand Cohn was also the first to formulate a scheme for the taxonomic classification of bacteria.

Louis Pasteur (1822-1895) and Robert Koch (1843-1910) were contemporaries of Cohn's and are often considered to be the founders of medical microbiology. Pasteur is most famous for his series of experiments designed to disprove the then widely held theory of spontaneous generation, thereby solidifying microbiology's identity as a biological science. Pasteur also designed methods for food preservation (pasteurization) and vaccines against several diseases such as anthrax, fowl cholera and rabies. Robert Koch is best known for his contributions to the germ theory of disease, proving that specific diseases were caused by specific pathogenic microorganisms. He developed a series of criteria that have become known as the Koch's postulates. Koch was one of the first scientists to focus on the isolation of bacteria in pure culture resulting in his description of several novel bacteria including *Mycobacterium tuberculosis*, the causative agent of tuberculosis.

While Louis Pasteur and Robert Koch are often considered the founders of microbiology, their work did not accurately reflect the true diversity of the microbial world because of their exclusive focus on microorganisms having medical relevance. It was not until the work of Martinus Beijerinck (1851-1931) and Sergei Winogradsky (1856-1953), the founders of general microbiology (an older term encompassing aspects of microbial physiology, diversity and ecology), that the true breadth of microbiology was revealed. Martinus Beijerinck made two major contributions to microbiology: the discovery of viruses and the development of enrichment culture techniques. While his work on the Tobacco Mosaic Virus established the basic principles of virology, it was his development of enrichment culturing that had the most immediate impact on microbiology by allowing for the cultivation of a wide range of microbes with wildly different physiologies. Sergei Winogradsky was the first to develop the concept of chemolithotrophy and to thereby reveal the essential role played by microorganisms in geochemical processes. He was responsible for the first isolation and description of both nitrifying and nitrogen-fixing bacteria.

Types of Microbiology

The field of microbiology can be generally divided into several subdisciplines:

- Microbial physiology: The study of how the microbial cell functions biochemically. Includes the study of microbial growth, microbial metabolism and microbial cell structure.
- Microbial genetics: The study of how genes are organised and regulated in microbes in relation to their cellular functions. Closely related to the field of molecular biology.

- Medical microbiology: The study of the role of microbes in human illness. Includes the study of microbial pathogenesis and epidemiology and is related to the study of disease pathology and immunology.
- Veterinary microbiology: The study of the role in microbes in veterinary medicine.
- Environmental microbiology: The study of the function and diversity of microbes in their natural environments. Includes the study of microbial ecology, microbially-mediated nutrient cycling, geomicrobiology, microbial diversity and bioremediation. Characterisation of key bacterial habitats such as the rhizosphere and phyllosphere.
- Evolutionary microbiology: The study of the evolution of microbes. Includes the study of bacterial systematics and taxonomy.
- Industrial microbiology: The exploitation of microbes for use in industrial processes. Examples include industrial fermentation and wastewater treatment. Closely linked to the biotechnology industry. This field also includes brewing, an important application of microbiology.
- Aeromicrobiology: The study of airborne microorganisms.
- Food Microbiology: The study of microorganisms causing food spoilage.
- Pharmaceutical microbiology: the study of microorganisms causing pharmaceutical contamination and spoillage.

Benefits of Microbiology

While microbes are often viewed negatively due to their association with many human illnesses, microbes are also responsible for many beneficial processes such as industrial fermentation (e.g. the production of alcohol and dairy products), antibiotic production and as vehicles for cloning in higher organisms such as plants. Scientists have also exploited their knowledge of microbes to produce biotechnologically important enzymes such as Taq polymerase, reporter genes for use in other genetic systems and novel molecular biology techniques such as the yeast two-hybrid system.

Fungal Biotechnology

Fungi in Agricultural Biotechnology

An area of agricultural biotechnology in which fungi show considerable potential for the future is the biological control of pathogenic fungi, insects & weeds. Such as biological approach to pest control is attractive because it provides an alternative to chemical pesticides which are increasingly perceived as being harmful to the environment. The second area of agricultural biotechnology where fungi are likely to be important is as inoculates or biofertilizers for use in increasing crop production.

Control of Fungal Plant Pathogens by Mycofungicides

The species of the genus Trichoderma have become the most widely investigated of all potential mycofungicides. Modern approaches to the use of this fungus to control pathogenic fungi have largely been based on the direct use of inoculants.

Spares of Trichoderma species can be readily & economically produced in the laboratory. But, the fact that no spectacular success has been achieved following the use of Trichoderma as an agent to control fungal pathogen, suggests that either members of this genus are not particularly efficient under field conditions, or more probably that we are not using them to best advantage. Safe & effective biocontrol results have been achieved with Trichoderma under very specialized conditions.

Biological Control

The use of chemical pesticides, has led to dramatic improvements in the production of crop plants. Since such improvements provided a reliable supply of cheap food they were initially welcomed of late, consumers are becoming increasingly concerned both about food quality & of the real & imagined effects of modern farming methods on the natural environment. While many of these fears have been exaggerated, there is neverthless a developing concensus that modern petro-chemical based farming is ultimately non-sustainable, as a result, more ecological approaches to food production are now being researched.

Pesticides were originally based on toxic heavy metals such as arsenic, mercury, lead or copper. Modern pesticides, however, are organic compounds, with a high degree of specificity towards their target organism. They also generally exhibit low over all toxicity & have little immediate impact on the environment. It is possible that the long term effects of these compounds might be subtly detrimental to soil fertility. Despite the positive results of the use of modern pesticides, concern continues to be expressed about the wisdom of using large quantities of chemicals in the environment. These fears have led biotechnologists to examine alternatives to chemical pesticides as a means of controlling agricultural pests. The most obvious, and apparently environment–friendly, alternative to pesticides is to use naturally occuring biological approaches to the control of pest populations.

Since all pests have natural antagonists, biological control should in theory be relatively straightforward. In practice, successful biological control is extremely difficult to achieve.

Fungi possess a number of characteristics that make them potentially ideal biocontrol agents. Firstly, many saprophytic species antagonise, representatives of all the pest organisms, including plant pathogenic fungi, weeds & insects secondly, fungi can be readily grown in culture so that large quantities can be economically produced for release, mainly as spores or mycelial fragments, into the environment. These inoculants then germinate or grow to produce active mycelium which can parasitise or otherwise inhibit the pest without damaging the non-target organisms. Fungi also survive for relatively long periods as resting bodies, and can then germinate to grow & control the target population thereby making continual reinoculation with the biocontrol agent unnecessary.

Use of Fungi to Control Inset Pest

Over 400 species of fungi attack insects & mites, so there is great potential for the use of these organisms as biological insecticides. As insect biocontrol agents, fungi are markedly superior to other micro-organisms because they are generally non-specific in their action & are useful against a wide range of insect pests. On the debit side, however, their use is limited by the restricted range of humidities in which they can grow,. An absolute humidity in which they can grow, an absolute humidity greater than 90% for a relatively long period is needed to enable fungi to grow & infect insect hosts. Another disadvantage is that fungi rarely, if ever, kill insects rapidly in the manner that we have come to associate with chemical insecticides.

Most of the so-called entomopathogenic fungi are phycomycetes & Deuteromycetes. Spores of these fungi attack either the external or gut cuticle of their insect hosts. They then germinate & hyphae penetrate the haemocoel. Death may result from the production by the fungus of a toxin, or following the direct utilization of the body fluids. After the insect body reserves have been used up, the mycelium then dies from lave of substrate & sporulates, the spare mass then acts as a focus of future infection.

Insecticidal toxins produced by fungi are non-enzymic, low molecular weight products, which can kill insects when present even at low concentrations. The best examples of the use of fungi to control insects are provided by species of Beauveria & Metarhizium, fungi which respectively cause white & green muscardine disease of insects.

The ideal mycoinsecticide should rapidly penetrate the cuticle of the insect, produce toxins to induce rapid paralysis, rapidly colonise the insects body & then sporulate under most environmental conditions. An obvious problem is that fungi used to control insects may be killed by fungicides sprayed on insect–infested crops.

Principal Deuteromycete Fungal Candidates for Arthropad

	Species	*Main Target Pests*
1.	Aschersonia aleyrodis	White fly
2.	Beauveria bassiana	Colorado beetle
3.	Beauveria brongniartii	Cockchafer
4.	Hirsutella thompsonii	Rust mites
5.	Metarhizum anisopliae	Beetles,bugs

6.	Nomuraea rileyi	Caterpillars
7.	Verticillium lecanii	Aphids, Whitefly

Commercial Production of Fungal Insecticies

Fungi can be readily produced for use as fungicides, as they are easy to grow on a variety of substrates, including agar-based media, cereal grains & a variety processed & unprocessed wastes; cereal grains are generally considered the best & most economical substrates for this purpose.

Product self-life is a major limitation of fungal biocontrol agents. A period of 18 months survival at 20° C is thought to be optimal, although production methods in current use can only achieve survival rates of around 3-6 months at storage temperatures of 40C Biocontrol fungi are usually mixed or formulated with substrates to increase their survival, in some cases colours or chemicals may be added to attract the insect host to the fungal pathogen.

Use of Fungi to Control Nematodes

Fungi that parasitise nematodes (nematophagus fungi) can be divided into nematode trapping fungi, endoparasitic species & fungi that parasitise nematode eggs. Nematode trapping fungi capture nematodes with specialised structures such as constructive & non-constructive rings, adhesive knobs or lastly by producing an adhesive material along the entire mycelial surface. Endoparasitic nematophagous fungi line in soils where they produce adhesive spores. These become attached to body of the nematode, on germination, a germ tube enters the body where it grows & consumes the host. Egg parasites, as their name suggests, are nematophagous fungi that parasitise the eggs of nematodes.

The following approaches have been used in an attempt of use fungi to control nematodes:

1. Exploitation of naturally supressive soils.
2. Soil amendments to encourage indigenous bio-control agents.
3. Inoculation of soils with selected strains of bacterial & fungi.
4. The use of microbal toxins & enzymes

1 Supressive Soils

Certain soils are naturally supressive to nemotodes. Fungi implicated in naturally supressive soils include Nematopthasa gynophila, Verticillium Chlamydosporium & Dactylella Oviparasitica.

2. Soil Amendments

A wide range of amendments have been found to be effective in controlling soil nematode populations. These generally involve breakdown products of added organic materials, including fatty acids & ammonia which together with an increase in natural antagonists prove effective in reducing nematode populations. Chitin amendment has been particularly extensively studied as a means of increasing populations of specific nematods antagonists.

3. Enzymes & Toxins

The direct application for enzymes, such as chitinase from egg parasitising fungi & collagenase from nematode — trapping species, to soils may also be useful means of controlling nematodes. Potentially useful nematocidal toxins are produced by fungi such as the oyster mushroom, pleurotus ostreatus.

4. Application of Selected Fungi

To date, Paecilomyces lilacinus is the only fungus that has been widely tested for nematode control under field conditions.

Use of Fungi to Produce Chemical Pesticides & Bio-fertilizers

Biologically active secondary metabolitises produced by fungi are being evaluated as potential pesticides, particularly for use in controlling plant growth. These compounds have the advantage over convential pesticides in being effective at very low concentrations while proving essentially non-persistant & harmless to the environment.

Use of Fungal Inoculants to Improve Crop Growth

Many of soil bacteria & fungi can liberate plant–available phosphorus from insoluble phosphates. This ability is usually associated with the production of organic acids & chelating agents, including citric acid & 2 — cetogluconic acid. Insoluble phosphates are found in abundance throughout the world, mainly as rock phosphate. Since rapidly available phosphate is desirable, the release of phosphorous, from these materials by natural weathering is far too slow to be a general benefit in agriculture. The rate of such phosphate release can, however, be stimulated by inoculating a microbial phosphate solubiliser into the soil. This could be a heterophilic bacterium, a member of the genus, Thiobacillus, or a phosphate solubilising fungus. Since fungi are heterotrophs, a nutrient source has to be added together with the inoculant & insoluble phosphate. Suitable carbon sources include molasses & sugarbeet bagasse.

Use of Fungi to Control Weeds

Weeds, in the widest sense of the world, include algae, ferns, annual & perennial plants, and also herbaceous or woody species. Fungi that are potential mycoherbicides can be divided into facultative parasites, which can be grown in the laboratory, and obligate parasites which at present cannot. Two approaches to used control using fungi have been developed (i) the inundative or mycoherbicide approach, and (ii) the classical biocontrol method. In the former approach, massive inoculations of fungi are applied at the optimum time for infection. The biocontrol agent then persists & controls the weed population. The classical method of biocontrol is achieved over a longer time period & depends upon the gradual increase in disease, the long term development of the biocontrol agent means that significant weed control may take months or even years.

Mycorrhizal Fungi as Inoculants for Use in Improving Crop Growth

Mycorrhizas are symbiotic associations between soil fungi & most higher plants. It was soon recongnized that mycorrhizal infection can often greatly increase the rate of uptake of nutrients such as nitrogen & phosphorus from nutrient-deficient soils. This has led to the view that the inoculation of mycorrhizal fungi into soils should lead to an increase in the uptake of these essential plant nutrients. Two types of mycorrhiza have been recongnized, the endotropic or vesicular -arbuscular mycorrhiza (VAM), & the ectotrophic type. In VAM the fungal partner is restricted to the cells of the plant cortex where it grows within & without the cells, invaginating the host cells at intervals to form a dichotomously branched structure called the arbuscle, thought to be the site of nutrient exchange between plant & fungus. The fungal partner appears to have no independent existence is soil. Neither is the interaction specific, since a single species of fungus can infect a wide range of plant, including most crop species. It is for this reason that VAMS are economically the most important type of mycorrhizal association. In ectoirophic mycorrhizas, the fungal partner forms a tight sheeth around the plant root & from this sheath hyphae grow into the outer cortex to form a network called the Hartig net. Ectotrophic mycorrhizas, unlike VAMS, tend to be non-specific. The fungal partner is readily grown in culture, and forms symbioses with trees in temperate forests. Vesicular-arbuscular mycorrhizas can directly enhance the uptake by plants of essential nutrients such as phosphorus, copper & iron on the other hand, Zinc & manganese uptake may be reduced, so that the mycorrhizal association many protect some plants from the toxic effects of the larger amounts of these elements. Ectotrophic mycorrhizas also show enhanced uptake of phosphorus, and by mineralising organic nitrogen make nitrogen available for plant use. They may also protect their plant hosts from heavy metals & attack by pathogens, and they may also help increase the uptake of water from soil to plants.

Some Biotechnology Important Fungi

Filamentous Species

Most of the filamentous fungi that are important in biotechnology are members of the important in biotechnology are members of the Deuteromycetes (Deuteromycotina).

Examples include Aspergillus niger, pencillum notatum — chrysogenum & Trichoderma Viridae. Basidiomycetes white rot fungi such as phanerochaete Chrysosporum are becoming increasingly important in environmental biotechnology because they are able to metabolise a variety of organic chemicals, many of which are pollutants.

Aspergillus SPP

There are about 200 species of Aspergillus, found throughout the world growing on a vast array of substrates. Aspergillus niger, as its specific name suggests, is a black fungi which is commonly called 'black mould'. Many Aspergilli contaminate foods, producing toxic products; A flavours for example, produce the mycotoxin alfatoxin. Several species also grow as contaminants on leather & cloth. Aspergillus species are important in medicine, causing disease of the internal organs. Aspergillus fumigatus accounts for nearly all of these infections. Aspergillus species produce numerous extracellular enzymes, many of which are put to good use in biotechnology. Aspergillus niger is particularly important in the manufacture of citric acid & gluconic acid, both finding wide application in food industry.

The asexual stage of Aspergillus is the one most often encountered by biotechnologists.

PE Nicillium SPP

Penicillin species are as widespread & cosmopolitan as the aspergilli. They are frequently referred to as green or blue moulds & are often found contaminating citrus fruits or causing decay on refrigerated cheese & other foodstuffs. Various species are also pathogenic to citrus fruits. Like members of the genus Aspergillus, penicillin species also attack leather & fabrics. The production of penicillin is without doubt the most significant use of this fungus. The only other therapeutically useful antibiotic produced by members of this genus is griseofulvin, which is used to treat fungal infections of the skin, including athelet's foot. Penicillium species also produce organic acids & flavours cheeses.

Trichoderma SPP

Trichoderma species are ubiquitous soil fungi which produce white, yellow or green colonies when cultured. Trichoderma species are used to produce cellulases. They are particularly effective as antagonists of the growth of other fungi, many of them plant pathogens, with the result that trichoderma species are important biocontrol agents.

Phanerochaete Chrysosporium

It is a member of the Basidiomycetes but it is the filamentous stage rather than the sporophore that is used in biotechnology. It belongs to a group called the white rot fungi, a name that emphasises the importance of these fungi as agents of wood decay. It is noteworthy for its ability to produce non-specific ligninases which can be used to degrade pollutants both in liquid effluents & in soils.

Yeasts

Yeasts play a fundamentally important role in biotechnology. The classic yeasts are unicellular & are slimy in appearance when growing on media often resembling bacterial colonies. Saccharomyces species are widely used in the food industry in bread production & in the fermentation of alcoholic beverages. Yeasts have also been used as source of single-cell protein, and to produce alcohol for industrial purposes. Recent developments in the molecular biology of yeasts have significantly expanded the biotechnological potential of these fungi.

Industrial Applications of Yeasts Organism Use

1. Saccharomyces Cerevisiae Brewing, baking, fuel, alcohol, wine
2. Saccharomyces uvarum Brewing, melibiase
3. Ashbya gossypii, Eremothecium ashbyii Riboflavin
4. Rhadotorula Carotene
5. Hansenula, Pichia SCP Via methanol
6. Yarrowia Lypolytica Citric acid
7. Saccharomycopsis, Aureobasidium D-Gluconic acid

Asaricus SPP

Agarius spp. are almost exclusively used in biotechnology for the production of mushrooms i.e. as a protein rich foodstuff. The commercial species used in mushroom growing in almost exclusively A bisporus.

Claviceps Purpurea

It is an Ascomycete notable for producing sclerotia, called ergot, from which a number of valuable, pharmacologically useful alkaloids are extracted.

Biocontrol of Bacteria & Phytopathogenic Fungi

Despite the many achievements of modern agriculture, certain cultural practices have actually enhanced the destructive potential of diseases. These practices include use of genetically similar crop plants in continuous monoculture, use of plant cultivars susceptible to pathogens, and use of nitrogenous fertilizers at concentrations that enhance disease susceptibly. Plant disease control, therefore, has now become heavily dependent on fungicides to combat the wide variety of fungal diseases that threaten agricultural crops. A land-mark study published by the U.S. Environment Protection Agency (EPA) indicates that, in the U.S. alone, 3000-6000 cancer cases are induced annually by pesticide residues on foods, and another 50-100 by exposure to pesticides during application. This type of findings have made the governments of many countries increasingly aware of the drawbacks of many chemical pesticides, in terms of their effect on the environment, as well as on the growers & consumers of agricultural products. Studies aimed at replacing pesticides with environmentally safer methods are safer methods are currently being conducted at many research centres. The heightened scientific interest is biological control of plant pathogens is partly a response to growing public concerns over chemical pesticides.

Biological control is a potent means of reducing the damage caused by plant pathogens. Commercialized systems for the biological control of plant diseases are few. Although intensive activity is currently being geared towards the introduction of an increasing number of biocontrol agents into the market. The performance of a bio-control agents into the market. The performance of a biocontrol agent cannot be expected to equal that of an excellent fungicide; although some biocontrol agents have been reported to be as effective as fungicide control.

Potential agents for biocontrol activity are rhizosphere-competent fungi & bacteria which, in addition to their antagonistic activity are capable of inducing growth responses by either controlling minor pathogens or by producing growth-stimulating factors.

Before biocontrol can become important component of plant disease management, it must be effective, reliable, consistent and economical to meet these criteria, superior strains, together with delivery systems that enhance biocontrol activity, must be developed. Existing biological control attributes can be enhanced by improving existing, known biological agents, with genetic manipulation. Genetic manipulations of biocontrol agents not only can enhance their activity, but also can expand their spectrum.

The growing interest in biocontrol with micro-organisms is also a response to the new tools of biotechnology plants and micro-organisms can now be manipulated to deliver the same mechanism of biological control, as has been done for the production of the delta endotoxin encoding gene transferred from Bacillus thuringiensis to plants to control insect pests. We can now think of micro-organisms with inhibitary activity against plant pathogens as potential sources of genes for disease resistance.

The successful control by biological means in the phylophane that have been reported involve mainly rusts powdery mildews and diseases caused by following genera of pathogens : Alternaria, Epicoccum, Sclerotinia, Spetoria, Drechisera, Venturia, Plasmopara, Erwinia and pseudomonas. Good soil biocontrol systems have been reported for species of Fusarium, Sclerotium, Scierotinia, Pythium and Rhizoctonia. The following biocontrol agents have already been registered; Agrobacterium radiobactor against crown gall (USA, Australia, NZ); Bacillus subtilis for growth enhancement (USA); Pseudomonas fluorescens against bacterial blotch (Australia); Pseudomonas fluorescens for seedling diseases (USA); Peniophora gigantea against Fommes annosus (UK); Pythium Oligandrum against Pythium spp. (USSR); Trichoderma viride against timber pathogens (Europe); Trichoderma spp. For root diseases (USSR); Fusarium oxysporium against Fusarium oxysporum (Japan); Trichoderma harzianum against root diseases

(USA); Gliocadium virens for seedling diseases (USA); Trichoderma harzianum/ Polysporum against wood decay (USA).

Mechanism of Biological Control of Plant Diseases.

Induced Resistance and Cross-protection.

Induced resistance is a plant response to challenge by microorganisms or abiotic agents such that following the inducing challenge de novo resistance to pathogens is shown in normally suceptible plants. Both localized and systemic induced resistance are nonspecific and can act against a whole range of pathogens, but whereas localized resistance occurs in many plant species, systemic resistance is limited to some plants. Cross-protection differs from induced resistance in that, following inoculation with avirulent strains of pathogens or other microorganisms, both inducing microorganisms and challenge pathogens occur on or within the protected tissue.

The most commonly reported examples of cross-protection involving fungi are probably those used against vascular wilts. Inoculation with nonpathogenic formae speciales of Fusarium and vernullum species, or with other fungi or bacteria, all have shown different levels of cross-protection.

Hypovirulence.

Hypovirulence is a term used to describe reduced virulence found in some strains of pathogens. This phenomenon was first observed in Cryphonectria (Endothia) parasitica (chestnut blight fungus) on European castanea sativa in Italy, where naturally occuring hypovirulent strains were able to reduce the effect of virulent ones. These slower growing hypovirulent strains contain a single cytoplasmic element of double-stranded RNA (ds RNA) similar to that found in mycoviruses, that was transmitted by anastomosis in compatible strains through natural virulent populations of C. Parasitica..

Hypovirulence has also been reported in many other pathogens, including Rhizoctonia Solani, Gaeumannomyces gramini var. tritici ophiostoma ulmi, but the transmissible elements responsible for hypovirulence or reduced vigor of the fungi are subjected to debate and may be due to ds RNAs, Plasmids, or viruses.

Competition.

Competition occurs between micro-organisms when space or nutrients (i.e. carbon, nitrogen and iron) are limiting, and its role in the biocontrol of plant pathogens has been studied for many years, with special emphasis on bacterial biocontrol agents. An important attribute of a successful rhizosphere biocontrol agent would be the ability to remain at high population density on the root surface, providing protection of the whole root for the duration of its life. Mycorrhizal fungi can also be considered to act as a sophasticated form of competition or cross-protection, decreasing the incidence of root disease.

Antibiosis

The production of antibiotics by actinomycetes, bacteria and fungi is very simply demonstrated in vivo. Numerous agar plate tests have been developed to detect volatile and non-volatile antibiotic production by putative biocontrol agents and to quantity their effects on pathogens. In general, however, the role of antibiotic production in biological control in vitro remains unproved. Species of Gliocadium and Trichoderma are well-known biological control agents that produce a range of antibiotics that are active against pathogens in vitro and, consequently, antibiotic production has commonly been suggested as a more of action for these fungi. Within bacterial biocontrol agents several species of the genus, Pseudomonas produce antibiotics involved in their ability to control plant pathogens.

Mycroparasitism :

Mycoparasitism occurs when one fungus exists in intimate association with another from which it derives some or all its nutrients while conferring no benefit in return. Biotrophic mycoparasites have a persistant contact with or occupation of living cells, whereas necrotrophic mycoparasites kill the host cells, often in advance of contact & penetration. Mycoparasitism is a commonly observed phenomenon in vitro & in vivo, & its mode of action & its involvement in biological disease control has been reviewed.

The most common example of mycoparasitism is that of Trichoderma SSP. Which attack a great variety of phytopathogenic fungi responsible for the most important diseases suffered by crops of major economic importance worldwide.

Biocontrol of Airborne Diseases :

Many naturally occuring microorganisms have been used to control diseases on the aerial surfaces of plants. The most common bacterial species that have been used for the control of diseases in the phylloshpere include Pseudomonas syringae, P. fluorescens, P. cepacia, Erwinia herbicola, and Bacillus subtilis, Fungal genera that have been used for the control of air borne diseases include Trichoderma Ampelomyces, and the yeasts Tilletiopsis & sporobolomyces.

Phytopathogenic bacteria possess several genes that encode phenotypes that allow them to parasitize plants & overcome defense responses elicited by the plant. In addition, phytopathogenic bacterial possess pathogenicity genes such as hrp. Isogenic avirulent mutants can be produced by insertional inactivation of genes involved in pathogenicity. Nonpathogenic mutants of Erwinia amylovora, produced by transposon mutagenesis, have also been used in the biological control of fire blight.

Antibiosis has been proposed as the mechanism of control of serveral bacterial & fungal diseases in the phyllosphere. Molecular biology techniques could be used to enhance the efficacy of biocontrol agents that use antibiosis as a more of action.

Biocontrol agents must normally achieve a high population in the phyloshpere to control other strains, but colonization by the agent may be reduced by competition with the indigenous microflora. Integration of chemical pesticides & biocontrol agents have been reported with Trichoderma spp. & P. syringae pv. Biocontrol agents tolerant to specific pesticides could be constructed using molecular techniques. Resistance to the fungicide benomyl is conferred by a single amino acid substitution in one of the B-tubulins of Trichoderma viridae. The corresponding gene thereby producing a biological control agent that could be applied simultaneously or in alternation with the Fungicide.

Biocontrol of Soil Borne Disease

Chemical control of soil borne plant diseases is frequently ineffective because of the physical & chemical heterogeneity of the soil, which may prevent effective concentrations of the chemical from reaching the pathogen. Biological control agents colonize the rhizosphere, the site requiring protection & leave no toxic residues, as opposed to chemicals.

Micro organisms have been used extensively for the biological control of soilborne plant diseases as well as for promoting plant growth. Fluroscent pseudomonas are the most frequently used bacteria for biological control & plant growth promotion, but bacillus & streptomyces species have also been commonly used. Trichoderma, Gliocadium, and coniothyrium are the most commonly used fungal biocontrol agents. Perhaps the most successful biocontrol agent of a soilborne pathogen is Agrobacterium radiobactor strain K84, used against crown gall disease caused by A-tumefaciens:

Competition as a mechanism of biological control has been exploited with soil borne Plant pathogens as with the pathogens on the phylloplane. Naturally occuring, nonpathogenic strains of Fusarium Oxysporium have been used to control wilt diseases caused by pathogenic Fusarium Spp. Molecular techniques have been used to remove various delterious traits of soilborne Phytopathogenic bacteria to construct a competitive antagonist of the pathogen.

Molecular techniques have also facilitated the introduction of beneficial traits into rhizosphere competent organisms to produce potential biocontrol agents. Chitin & b - (1,3) — glucan are the two major structural components of many plant Pathogenic fungi, except by oomycetes, which contain cellulose in their cell wall & no appreciable levels of chitin. Biological control of some soilborne fungal diseases has been correlated with chitinase production, bacterial producing chitinases or glucanases exhibit antaganosm in vitro against fungi. A recombinant Escherichia coli expressing the chi A gene from S marcescens was effective in reducing disease incidence caused by screrotium rolfsii & Rhizoctonia solani. In other studies, chitinase genes from S. marcescens have been expressed in Pseudomonas spp. & the plant symbiont Rhizobium meliloti. The modified Pseudomonas strain controlled the pathogen F. Oxysporium f. species rodelens & Gauemannomyces graminis var. tritici.

The Trichoderma System

Trichoderma spp. act against a range of economically important aerial & soilborne plant pathogens. They have been used in the field & greenhouse against silver leaf on plum, peach & nectarine; Dutch elm disease on elm's honey fungus (Armillaria mellea) on a range of tree species; and against rots on a wide range of crops, caused by fusarium, Rhizoctonia, and pythium, and sclerotium forming pathogens such as Sclerotinia & Sclerotium. In many, experiments, showing successful biological control, the antagonistic Trichoderma was mycoparasite.

Mechanism of Action :

Form recent work, it appears that mycoparasitism is a complex process, including several successive steps. The first detectable interaction shows that the hyphae of the mycoparasite grows directly towards its host. This phenomenon appears a chemotropic growth of Trichoderma in response to some stumuli in the hosts's hyphae or toward a gradient of chemicals produces by the host.

When the mycoparasite reaches the host, its hyphae often coil around it or are attached to it by forming hook like structures. In this respect, production of appressoria at the tips of short branches has been described for T. hamatum & T. harzianum. The interaction of Trichoderma with its host is specific. The possible role of agglutinins in the recongnition process determining the fungal specificity has been recently examined. Indeed, recognition between T. harzianum & two of its major hosts, R. solani & S. rolfsii, was controlled by two different lectins present on the host hyphae. R. solani carries a lectin that binds to galactose & fucose residues on the Trichoderma cell walls. This lectin agglutinates conidia of a mycoparasitic strain of T. harzianum, but did not agglutinate two non-parasitic strains. This agglutinin may play a role in prey recognition by the predator Moreover, because it does not distinguish among biological variants of the pathogen, it enables the Trichoderma species to attack different R. Solani isolates. The activity of a second lectin isolated from S. rolfsii was inhibited by d-glucose or d-mannose residues, apparently present on the cell walls of T. harzianum.

Following these interactions the mycoparasite sometime penetrates the host mycelium, apparently by partially degrading its cell wall Microscopic observations led to the suggestion that Trichoderma spp. Produced & secreted mycolytic enzymes responsible for the partial degradation of the host's cell wall. The Complexing & diversity of the chitinolitic system of T. harzianum involves the complementary modes of action of six enzymes, all of which might be required for maximum efficiency against a broad spectrum of chitin-containing plant pathogenic fungi.

The level of hydrolytic enzymes produced differs from host-parasite interaction analyzed. This phenomenon correlates with the ability of each Trichoderma isolate to cotnrol a specific pathogen. It is considered that mycoparasitism is one of the main mechanisms involved in the antagonism of Trichoderma as a biocontrol agent.

The process apparently includes:

1. chemotropic growth of Trichoderma,
2. recognition of the host by the mycoparasite
3. secretion of extracellular enzymes,
4. hyphae penetration, and
5. lysis of the host.

The involvement of volatile & nonvolatile antibiotics in the antagonism by Trichoderma has been proposed. Indeed some isolates of Trichoderma excrete growth-inhibitary substances. Thus, the biocontrol ability of Trichoderma strains is most likely conferred by more than one exclusive mechanism. In fact, it seems advantageous for a biocontrol agent to supress a plant pathogen using multiple mechanisms.

Agricultural Biotechnology

Every year, millions of people worldwide die from starvation and nutritional deficiencies. Although these challenges remain significant, we are making some progress. Agricultural technologies have increased caloric intake on a per-capita basis in nearly every nation of the world development — particularly in the area of agricultural biotechnology — is crucial if we want to further reduce malnutrition and starvation and meet the needs of growing populations.

Unfortunately, misperceptions about biotechnology have led to activist movements to slow down, if not eliminate, this critical technology via government regulation. In the United States, the advancement of biotech crops has slowed down because of attempts to regulate them differently than conventionally grown foods. In Europe, antibiotech movements are much stronger, and international bureaucrats are pushing various international agreements and trade restrictions. While Western na- tions may be able to absorb the costs of such policies, people in the developing world will pay a higher price as many continue to suffer from starvation.

What Exactly Is Biotechnology?

For centuries breeders have improved plants through selective breeding and crossbreeding different types of plants. Agricultural biotechnology — also referred to as "genetic engineering" or "genetic modification" (GM) — is a more sophisticated and precise means of modifying plant genes. Instead of transferring thousands of genes as traditional crossbreeding does, biotechnology enables the transfer of only a few select plant genes. Breeders essentially cut the desirable traits out of one plant's DNA and paste them into another — a process called "gene splicing." In addition, undesirable traits can be targeted and removed. Using biotechnology, plant breeders can produce better crops.

Biotechnology's Critical Benefits

More affordable, higher quality food surely benefits people in developed nations, but those benefits pale in comparison to what biotech promises for the developing world.

- Biotechnology promises to reduce world hunger and disease by improving local productivity by adapting crops to local climates and soils; increasing yield by making plants stronger and more pest-resistant; making plants more nutritious by creating plants with higher vitamin and protein content; and making produce more affordable on the world market.
- For example, Cornell University researchers are developing a rice variety that yields 20 percent to 40 percent more food per acre and a tomato that produces 48 percent higher yields per acre.
- Biotechnology promises to play a critical role in reducing blindness caused by vitamin A deficiencies. Every year in Southeast Asia, one quarter of a million children go blind because of a lack of vitamin A in their diets. Worldwide, 124 million children suffer from vitamin A deficiency. Increasing the intake of vitamin A could prevent one to two million deaths per year and eliminate numerous health problems. Poverty and lack of infrastructure make the traditional methods of vitamin consumption (a varied diet or supplements) impossible. Agricultural biotechnology, however, can help to address the problem of childhood blindness and death by ensuring that people get this nutrient in the key staple of their diet, rice. Researchers have spliced into the genome of rice, the beta-carotene gene from a daffodil to create "golden rice" — rice high in beta-carotene, which the human body converts into vitamin A. This is among the most well-known examples of biotechnology's benefits. Countless other genetically engineered crop varieties are being developed to address similar problems of poor nutrition or productivity.

How Agricultural Biotechnology Improves Plant Safety

Many biotechnology skeptics contend that such engineering is unsafe because it is "unnatural." Ironically, biotechnology is actually an improved method for plant breeding. Most of what we eat today has already undergone gene manipulation under the traditional breeding methods. The traditional methods and biotechnology differ in that traditional methods are cruder because they do not give researchers control over all the genes they crossbreed. Some desirable characteristics might result with such methods, but undesirable ones might as well. It could take a long time of continual breeding to eliminate the most problematic traits and focus on the best ones.

Compared to these more "traditional" methods of hybridization, genetic engineering is a wonder of precision and accuracy. Genetic engineering enables scientists to pinpoint the genes that express a desirable characteristic in one plant and to insert those genes into the genome of another plant. Scientists mark the insertion point so that they know where these novel genes are located in the genome of the new plant. While there are several different methods of genetically engineering plants, these processes transfer far fewer genes than do "traditional" methods of plant breeding.

Agricultural Biotechnology Regulation

For traditionally bred plants, regulators rely on plant breeders to conduct appropriate safety testing and to be the first line of defense against genetic alterations that might prove dangerous. The same should be true for agricultural products developed using biotechnology. There is no reason for regulators to treat biotech plants any differently than traditionally bred plants, particularly given the fact that biotechnology provides greater control over gene manipulation.

Despite fears about gene manipulation, traditional crossbreeding has altered plant genes and improved the human diet for the past 30 years. In fact, people worldwide safely consume these plants every day. Examples include wheat, corn, potatoes, tomatoes, and countless other staples of the American diet. In 1984, the biotechnology industry began experimenting with "gene-spliced plants" — the more advanced approach to gene manipulation that we now call "biotechnology."

Regulatory Scheme

At the time, the White House Office of Science and Technology Policy began crafting a framework wherein existing federal agencies would regulate genetically engineered organisms on the basis of their characteristics, not the method of production — thus wisely deferring to the scientific community's judgment that regulation ought to address the products, not the process of biotechnology. Engineered organisms would not require extra scrutiny simply because genetic engineering produced them (the process). Instead, they would be subject to heightened scrutiny only if the individual organisms expressed characteristics that posed some conceptually heightened risk (the product).

The federal government divided regulatory jurisdiction among agencies already involved in agricultural, food, and environmental regulation. These include the United States Department of Agriculture (USDA), the Environmental Protection Agency (EPA), and the Food and Drug Administration (FDA). While each of these agencies considers the characteristics of individual products in their regulation, only FDA followed the general scientific thinking that genetically engineered and non-genetically engineered products should be regulated similarly. Both USDA and EPA automatically subject all genetically engineered plants as a class to premarket approval requirements not ordinarily applied to conventionally bred plants.

Department of Agriculture

USDA regulates the release of all genetically engineered agricultural plants under statutes giving USDA's Animal and Plant Health Inspection Service (APHIS) the authority to regulate plants that may be or may become weeds or other nuisances — what the statutes call "plant pests." Although the rules apply in a general sense to novel or exotic varieties of both gene-spliced and conventional plants, APHIS typically only requires field testing of conventional plants that are new to a particular U.S. ecosystem (transplanted from another continent, for example).

However, all genetically engineered agricultural plants face a higher regulatory bar — making it more difficult to expand the use of biotechnology. Genetically engineered crops must be field tested under APHIS regulations prior to commercialization. For example, a new variety of corn produced with conventional hybridization requires no government-mandated field testing, but all new varieties of genetically engineered corn do, even though there is no logical reason for the regulatory disparity. For most genetically engineered plants, APHIS requires the company producing the plants to submit notice detailing the gene or genes that have been inserted, where the plant will be tested, and other relevant characteristics of the plant prior to receiving permission to conduct the field trials. Once the company completes field testing, APHIS reviews the results and makes a determination on whether or not the product should be "deregulated" and can be released into the market.

Environmental Protection Agency

EPA regulates plants that are modified to produce substances that act like pesticides: that is, substances used by a plant to protect itself from pests, such as insects, viruses, and fungi. EPA's proposed rule for these "plant-incorporated protectants" is not yet finalized. In the interim, however, EPA has regulated such plants by applying its proposed guidelines, which are functionally similar to rules for the registration of synthetic chemical pesticides. Again, biotech products face higher regulatory hurdles, even though plant-incorporated protectants developed through conventional breeding are exempt from these requirements.

Health and Human Services: Food and Drug Administration

FDA is responsible for ensuring that food items, including foods derived from genetically modified plants, are safe to eat. Following the general regulatory framework that emphasizes product regulation rather than process regulation, FDA rightly does not treat foods derived from genetically engineered plants to be inherently unsafe.

Food producers are not required to seek premarket approval from FDA unless there is a substantive reason to believe that the novel trait (or traits) in the food pose a safety question. The initial determination of safety is left to the producer, but FDA has encouraged producers to consult with agency scientists prior to marketing a food produced with bio-technology to ensure that the appropriate determination is made. Recently, FDA published a proposed rule that would require producers to notify the agency at least 120 days before marketing, 16 which may be a signal that FDA could abandon its more reasonable approach in the future.

In addition, FDA does not require labeling of foods derived from biotechnology unless the genetic insertions so alter the food that the common name no longer describes it adequately. Examples of this phenomenon would include such alterations of the food product that would raise a safety issue or change the product's nutritional composition or its storage or preparation characteristics.

Labeling Issues

Some activists, however, argue that the government should mandate the labeling of all genetically engineered foods. They assert that consumers have a "right to know" how their foods have been altered, and that a mandatory label would best allow consumers to choose between genetically engineered and conventional foods.

Biotechnology advocates have argued against mandatory labeling because such requirements raise food costs — something that mostly harms lower-income Americans and people on fixed budgets. Perhaps more important, while biotech products are not substantially different from other products, special labels would likely make consumers think these products were more dangerous. Hence, rather than serving educational or "right to know" purposes, such labels promise to simply confuse consumers.

A government-mandated label on all genetically engineered foods also would raise important First Amendment free speech issues. In 1996, the U.S. Court of Appeals, in *International Dairy Foods Association et al. v. Amestoy*, ruled unconstitutional a Vermont statute requiring the labeling of dairy products derived from cows treated with a genetically engineered growth hormone, noting that food labeling cannot be mandated simply because some people would like to have the information. "Absent ... some indication that this information bears on a reasonable concern for human health or safety or some other sufficiently substantial governmental concern, the manufacturers cannot be compelled to disclose it." In other words, to be constitutional, labeling mandates must be based in science and confined to requiring disclosure of information that is relevant to health or nutrition.

Furthermore, consumers need not rely on mandatory labeling of biotech foods to truly have a choice. Real-world examples show that market forces are fully capable of supplying information about process attributes (including kosher and organic production standards) that consumers truly demand. The same can be said about nonbiotech foods, and the FDA recently published proposed guidelines to assist producers in voluntarily labeling both genetically engineered and nongenetically engineered foods. Additionally, the USDA's newly published rule for organic certification necessarily excludes biotech products from organic food production. Consequently, consumers wishing to purchase nonbiotech foods need only look for certified organic products.

International Trade

While U.S. consumers do not appear to be strongly opposed to biotech foods (they, in fact, seem rather indifferent), a strong anti-biotechnology movement has arisen in several European countries over the last few years. The European Union (EU) has established strong restrictions on the commercial planting of genetically engineered plants, and European food processors and retailers are reluctant to import harvested agricultural products derived from biotechnology — thus jeopardizing the marketability of U.S. commodity grain exports. And the EU is now negotiating for strong restrictions on agricultural biotech products in important international agreements governing food safety and environmental protection. Very strong restrictions were included in the Biosafety Protocol, finalized in January 2000. Perhaps, more important, though, are the Codex Alimentarius Commission standards for food safety.

Without convincing scientific evidence that genetically engineered crop plants pose a height-ened environmental or human health risk, restrictions on agricultural biotech imports could be challenged under the General Agreement on Tariffs and Trade and the World Trade Organization. Thus the EU has resorted to justifying its decisions on the basis of a risk-management philosophy known as the "precautionary principle," and the EU advocates inclusion of the precautionary principle in international environmental and food safety agreements (such as the Codex Alimentarius), as well as within GATT itself. Its inclusion would give nations greater latitude in restricting imports of U.S. agricultural products.

No single definition exists for the precautionary principle, but its general meaning is that when an activity raises threats of harm to human health or the environment, regulatory measures should be taken to prevent or restrict the activity even if the risks have not been demonstrated scientifically. Although the EU asserts that the precautionary principle is an unbiased risk-management philosophy, critics have noted that its lack of definition and evidentiary standards makes it all too easy to abuse for the purpose of masking trade protectionism, and that its very approach to risk management is inherently flawed and may, in fact, increase net risk. Nevertheless, the precautionary principle has already been included in several international environmental treaties, including the Convention on Biological Diversity and the Biosafety Protocol.

Even as underdeveloped nations clamor for biotechnology applications, and as countries like China continue to experiment with and use agricultural biotechnology, opponents of agricultural biotechnology in the West, particularly Europe, attack it as an unnatural process that will destroy the world, not better it. They argue that biotechnology should be heavily regulated, if not banned. Already, however, genetically engineered plants are subject to strict regulatory oversight that is equal to or greater than that advocated by the vast majority of scientific specialists. Additional regulation will slow down research and development of genetically engineered crops, keep beneficial products off the market, and raise the cost of products that do make it to consumers. Furthermore, the inclusion of similar restrictions — or inclusion of the precautionary principle — in international agreements will greatly impact the international trade of agricultural goods and delay their introduction into the marketplace. Each of these problems could prevent this technology's benefits from being introduced to industrialized nations and, more important, the developing world.

Prospects and Limitations of Agricultural Biotechnologies: An Update

It is now generally that the practice of plant breeding will move forward as progress is made in knowledge and technology.

Breeders in the future will not only be forced to hone their traditional skills but will also have to integrate them with knowledge and experience to coincide with more recent advancements and technologies derived from cellular and molecular concepts and approaches. For worldwide crops such as potato, rice, maize and in the future wheat, as well as for high value vegetable crops, the future of breeders depends upon the use of a combination of breeding practices (traditional breeding, genetic engineering and tissue culture). In addition to an increasing reliance on breeding, agriculture will also depend on biocontrol to complement and to make possible the use of chemicals that are compatible with intelligent management of the natural resources essential for a sustainable yet highly productive agriculture.

The various chapters in this book document these points very convincingly and relatively exhaustively. Thus most aspects of agriculturally relevant biotechnology are covered. Plant tissue culture techniques not only provide essential ways for the clonal propagation of many agriculturally important crops (e.g woody plants ornamentals and vegetables) but they are also the basis for the production of transgenic plants and are at the forefront of recent studies in plant physiology, developmental biology and biochemistry. Spectacular progress in our knowledge of plant sciences, including those relevant to agriculture and breeding, can be expected from the integration of more traditional sciences with molecular and cellular concepts and techniques. A prime example is seen in the impact molecular genome analysis is already having on traditional breeding (via so called map-based breeding) but also on research and molecular breeding (e.g, via map based gene-cloning)

This rosy description of the future of genetic engineering is richly tempered in this book by a realistic assessment of the many problems that still need to be faced before what is predicted can be accomplished.

It was therefore wise to ask scientists well known for their "no-nonsence" approach such as Ingo Potrykus and his colleagues to write about genetic engineering of crop plants.

To quote from their introduction: " it has been possible to develop the state of the art of gene transfer to plants to a level at which many of the major crop plants are accessible to the technique and for which we have good reason to believe that there is no basic biological principle that will prevent gene transfer to a specific crop plant " Yet these authors clearly identify the considerable research and development that is still needed before one can, in many but not all instances, consider large scale applications.

A recurrent point made and well documented in various chapters of this book is that genetic engineering—because it is not limited by the natural barriers that ensure the maintenance of species and prevent transfer of genes between unrelated organisms—is an extremely powerful way to increase the genetic variability available to breeders.

The present rise in interest to acquire new knowledge via fundamental research can best be illustrated by reminding ourselves of recent progress in our understanding of the molecular physiology of development, of the mechanism of action of old and newly discovered phytohotmones and of growth signals used by microorganisms that are pathogens or symbionts of plants (such as for instance lipo-chitooligosaccharides). We can except to see more chapters on these topics in future books on biotechnology.

Biotechnology can contribute to making traditional agriculture more productive while at the same time more sustainable. It can also help achieve the goals of "non-food" and or cash crops agriculture as has been described in the various chapters dealing with crop improvement and metabolite production.

Hopefully sufficient attention will be paid to regional crops of importance by local populations in developing countries.

Finally topics such as "Biofertilization" and "Bioremediation" are convincingly dealt with Topics such as "Biotechnology of Farm Animals" and "Marine Biotechnology" as well as the unavoidable but all important "Legal and Public Aspects" round out the all inclusive title of Agricultural Biotechnologies.

Some topics have already come to blossom and illustrate well the potential of present day plant sciences. There is the breakthrough in the cloning and characterization of resistance genes along with significantly improved understanding of the nature of avirulence genes and elicitors. Many hypothetical models predict that resistance genes should be involved in the signal transduction pathways linking pathogen derived signals with the regulated expression of defense genes and of programmed cell death (hypersensitivity reaction). The first available data on resistance genes show that this concept may well be correct in its very general sense, even if all resistance genes do not code for elicitor receptors.

We may safely predict that in the near future we will know more details of the molecular mechanisms underlying plant-pathogen resistance mechanisms. Several elicitor receptors will soon be isolated and characterized along with genes involved in localized pathogen-induced programmed cell death. Plant breeders will make use of isolated resistance genes to produce crops with improved resistance to a variety of pathogens and pests.

The content of this book fully justifies the conclusion that Biotechnologies are bound to play an ever growing role in agriculture not only in the long term future, but also presently, and in the near future.

Fortunately the progress in applied plant biotechnology corresponds with and is in fact stimulating fundamental scientific progress. This is important because the real driving force behind this science is the success of applied plant biotechnology such as: 1) plant protection based on the expression of several different introduced genes coding for different principles acting on the same targets and thereby providing a possible solution to the problem of the emergence of resistant pathogens and pests, 2) new quality traits as well as future crops producing tailor-made non-food products (lipids, carbohydrates, and biodegradable thermoplastics). Only if plant biotechnology becomes a commercial as well as an environmental success will sufficient support be available for this fascinating research.

Plants and Agriculture

Human population has grown steadily since the advent of agriculture, nearly 12,000 years ago, which ensured a continuing and reliable supply of food. The most rapid growth occurred during the past two centuries during which time agriculture became highly industrialized, public health improved and there were increases in incomes and international trade in food grains. Despite the age-old human concerns about balancing population and food supplies and many population control measures the worlds population is projected to continue to grow into the early decades of the 21st century, stabilizing at 9-11 billion sometimes during 2030-2050. Much of this growth will occur in the already overpopulated underdeveloped and poorer regions of Africa, Asia and Latin America, which will be home to nearly 90% of the human population.

Food shortages were common during the early and middle years of the 20th century. The introduction of Green Revolution varieties of wheat, rice and maize during the mid-1960s reversed this trend and helped avoid major food shortages by keeping increases in food productivity slightly ahead of population growth. During this period, both China and India, the two most populous nations with chronic food shortages, became bet food exporters. Realistically however such increases in food productivity can not be sustained indefinitely. It is not surprising, therefore that increases in food productivity have begun to decline during the past few years. Furthermore, improved economic conditions in China and India have created greater demand for better and varied food products, particularly poultry and meat, which require greater supplies of feed grains. As a result, both China and India have now become net importers of food. World food reserves have declined from a high of 77 days to less that 50 days.

With the current trends of population growth and agricultural production the demand for food in the most populous parts of the World will double by the year 2025, and nearly triple by 2050, Increases in food productivity of this magnitude can not be brought about in such a short period of time by conventional breeding, especially when some of out most important crops are approaching the physiological limits of productivity, not by increasing the amount of arable land, which accounts for 97% of all food production in the world.. Arable land, which is finite and comprises about 3% of the earth's surface, is deteriorating and decreasing as a result of soil erosion, salinization, over cultivation, and acidification. These factors, combined with expected increases in population, will actually decrease the global per capita arable land from the current 0.28 hectare to 0.17 hectare by the year 2025. In addition, fresh water supplies, essential for modern high-input irrigation agriculture, are becoming limited by increased human and agricultural use, and polluted by agricultural run-off and widespread use of agrochemicals.

It is feared that the resulting food shortages in the overpopulated parts of the world during the 21st century may lead to widespread social, economic, and political unrest, making food security the single-most serious threat to international peace and security. At the same time it is well known that countries with efficient agricultural systems generally have high living standards, strong economies, lower rates of population growth and democratic forms of government Increasing food productivity in a sustainable manner will, therefore, not only provide adequate nutrition to the expanding humanity, but will reduce population growth, protect the environment, promote economic development and ensure social and political stability.

The challenge for the agricultural sector during the next few decades is therefore clear double food production by 2025, and triple it by 2050, on less per capita land, with less water, under increasingly challenging environmental conditions. The situation is further complicated by the fact that in spite of the heavy use of agrochemicals, modern agriculture still loses nearly 42% of crop productivity to competition with weeds and to pests and pathogens, and an additional 10-30% to post-harvest losses due to a variety of factors, especially in the developing countries where storage conditions are poor. During the 20th century traditional plant breeding has brought about enormous increases in crop productivity. However, plant improvement by hybridization is slow, and is restricted to a very small gene pool owing to natural barriers to crossability. Beginning in the early 1980s, advances in plant cell culture can genetic transformation have overcome these barriers by making it possible to transfer defined genes into all major food crops, including cereals, legumes, cassava, potato, and many vegetable and fruits. The entire global gene pool—whether it be plant, animal, bacterial or viral is available for utilization. The first genes that have been integrated into crop species provide resistance to non-selective and environment friendly herbicides, and many pests and pathogens. Increasingly large acreages of transgenic maize, soybean, potato, tomato and cotton are being commercially grown for human use and consumption. Carefully planned introduction of such crops on a worldwide scale would greatly help in reducing or even elimination the enormous crop losses attributed to weeds, pests and pathogens. The use of such crops will also have a beneficial effect on the environment by significantly reducing the use of agrochemicals. Other genes for improving crop productivity, and manipulating starch/protein/oil quality and quantity, resistance to environmental stresses such as temperature and drought are also being isolated and studied. In the foreseeable future, these will be used to produce second generation transgenic crops.

It is clear that during the next few decades a wide variety of transgenic crops will become integrated into agricultural systems in the industrialized countries, Introduction of such crops into the developing countries, which need them most, will be slow and largely through the efforts of multinational biotechnology companies, because most of the developing countries currently lack the scientific and industrial infrastructure to develop and introduce these technologies into their agriculture. In the long term however, much of the increase in food production to meet the dual challenge of population growth and food demand must occur in the developing countries. This will impact greatly on their overall economic development, which will help control the relentless increase in the population. It is critical, therefore that scientific and technical manpower and infrastructure be created in developing countries to take scientific and technical manpower and infrastructure be created in developing countries to take advantage of the remarkable advances in agricultural biotechnology. Assistance should be provided to the developing countries be international organizations such as the United Nations Educational Scientific and Cultural Organization

(UNESCO), the Food and Agriculture Organization of the United Nations (FAO) and the World Bank through its many international agricultural research centers, like those in Mexico and the Philippines, where Green Revolution originated, International agriculture in the latter half of the 20th century was dominated by the Green Revolution, Considering the power and potential of agricultural biotechnology, it is likely that the Gene Revolution will dominate the agriculture of the 21st century. Like the Green Revolution, it may help accelerate the rate of food productions, save lives from hunger, create livelihoods in rural households, save large tracts of land that would otherwise be needed for food production, reduce birth rates, ensure low food prices, stimulate broad-based economic growth, expand world trade, and help create a sustainable agricultural system for future generations. Finally, it should be understood that although plant breeding has long been, and will continue in the future, to be indispensable for plant improvement, it must now be complemented and supplemented by molecular breeding and genetic transformation, in order to establish a sustainable agricultural system for the 21st century.

Animal Biotechnology

Animal Biotechnology has developed rapidly from the early 1980s when the first transgenic mice and first in vitro produced bovine embryos occurred. Today animal breeding companies are using marker assisted selection to provide earlier and improved selection of breeding animals. Gene mapping efforts are annually identifying a large number of candidate genes for testing of usefulness in marker assisted selection.

The propagation of genetically valuable or transgenic animals is enhanced by in vitro production of embryos. There are already more than a dozen commercial companies selling or producing for customers in vitro produced embryos. Cloning of embryos of cattle and sheep has been shown to be possible, but low efficiencies have thus far prevented commercial use. Animal producing and genetic selection systems can be made more efficient by sexing the sperm used for in vitro production of embryos. The flow sorting of sperm cells to separate X from Y bearing cells has been successful in most species tested and in cattle and swine has resulted in offspring of the desired sex. While effective with in vitro embryos production, the sexed sperm are of less than normal viability and sperm numbers and not yet useful in artificial insemination or the freezing of sperm.

The artificial insemination and embryo transfer industries are today heavily dependent on use of cryopreserved sperm and embryos. However the cryopreservation of oocytes and the combination of some biotechnologies such as embryo biopsy and sexing with cryopreservation are still in the research stage. Transgenic cattle, sheep swine and poultry have been produced primarily by microinjections of DNA into pronuclei of eggs or by use of viral vectors. The intended use has been to change animal growth, disease resistance, wool production, or to produce new products in milk. The efficiencies thus far have been much lower than in mice. Therefore there has been little commercial use of transgenic technology except for the involvement of several companies in production of pharmaceutical products in milk.

The use of cultural embryonic stem cells and nuclear transfer to make offspring has resulted in calves and lambs. This technology offers promise of making large numbers of nearly identical offspring or more efficient gene transfer or selection. However thus far the technology is not sufficiently efficient for commercial use. Growth hormones and growth promoting factors are being developed for improved efficiency of animal growth. Growth hormones are commercially in use at present for promotion of milk production in cattle.

The anatomy and physiology of birds and the late stage of egg laying prevents the introduction of DNA into pronuclei as in manuals. Thus other methods such as oocytle microinjection of DNA, viral introduction of DNA and introduction of DNA into primordial germ cells are being used in avian gene transfer research. Oocyte microinjection, blastoderm transfection and primordial germ cell manipulation have all been successful in producing transgenic birds. The production of transgenic birds appears possible but is not yet in commercial use.

Patented applications of recombinant DNA biotechnology to animal agriculture became apparent in the early to mid 1980s with development of recombinant vaccines effective against calf scours and against rabies. Today numerous vaccines in use are produced by recombinant technology. Biotechnology has also revolutionized diagnostic techniques by introducing higher sensitivity and specificity. Monoclonal antibodies, DNA probes, Southern and Western blotting, DNA and RNA amplification through PCR, constitute as integral part of the diagnostic arsenal at the disposal of the advanced laboratories. Based on these techniques rapid animal-side field test used by the clinicians furnish results in real time, permitting immediate decisions as to the proper treatment, vaccination, or stamping-out policy. High technology based diagnosis is expected to become a routine in every diagnostic laboratory forming a sound basis for control and treatment of animal diseases.

In the short time of approximately fifteen years, animal biotechnology has moved from an unknown and unstudied science to a strong discipline of science and extensive commercial application. The science and its commercial use are rapidly progressing. The ultimate benefactor will be humans through improved food supply, nutrition and health care products available to humans around the world.

Agricultural Production Applications

Humans have always relied on plants and animals for food, shelter, clothing and fuel, and for thousands of years farmers have been changing them to better meet our evolving needs. Society's demand for resources provided by plants and animals will increase as the world's population grows. The global population, which numbered approximately 1.6 billion in 1900, has surged to more than 6 billion and is expected to reach 10 billion by 2030. The United Nations Food and Agriculture Organization estimates world food production will have to double on existing farmland if it is to keep pace with the anticipated population growth.

Biotechnology can help meet the ever-increasing need by increasing yields, decreasing crop inputs such as water and fertilizer, and providing pest control methods that are more compatible with the environment.

Crop Biotechnology

Farmers and plant breeders have relied for centuries on crossbreeding, hybridization and other genetic modification techniques to improve the yield and quality of food and fiber crops and to provide crops with built-in protection against insect pests, disease-causing organisms and harsh environmental conditions. Stone Age farmers selected plants with the best characteristics and saved their seeds for the next year's crops. By selectively sowing seeds from plants with preferred characteristics, the earliest agriculturists performed genetic modification to convert wild plants into domesticated crops long before the science of genetics was understood.

As our knowledge of plant genetics improved, we purposefully crossbred plants with desirable traits (or lacking undesirable characteristics) to produce offspring that combine the best traits of both parents. In today's world, virtually every crop plant grown commercially for food or fiber is a product of crossbreeding, hybridization or both. Unfortunately, these processes are often costly, time consuming, inefficient and subject to significant practical limitations. For example, producing corn with higher yields or natural resistance to certain insects takes dozens of generations of traditional crossbreeding, if it is possible at all.

The tools of biotechnology allow plant breeders to select single genes that produce desired traits and move them from one plant to another. The process is far more precise and selective than traditional breeding in which thousands of genes of unknown function are moved into our crops.

Biotechnology also removes the technical obstacles to moving genetic traits between plants and other organisms. This opens up a world of genetic traits to benefit food production. We can, for example, take a bacterium gene that yields a protein toxic to a disease-causing fungus and transfer it to a plant. The plant then produces the protein and is protected from the disease without the help of externally applied fungicides.

Improving Crop Production

The crop production and protection traits agricultural scientists are incorporating with biotechnology are the same traits they have incorporated through decades of crossbreeding and other genetic modification techniques: increased yields; resistance to diseases caused by bacteria, fungi and viruses; the ability to withstand harsh environmental conditions such as freezes and droughts; and resistance to pests such as insects, weeds and nematodes.

Natural Protection for Plants

Just as biotechnology allows us to make better use of the natural therapeutic compounds our bodies produce, it also provides us with more opportunities to partner with nature in plant agriculture.

Through science, we have discovered that plants, like animals, have built-in defense systems against insects and diseases, and we are searching for environmentally benign chemicals that trigger these natural defense mechanisms so plants can better protect themselves.

Biotechnology will also open up new avenues for working with nature by providing new biopesticides, such as microorganisms and fatty acid compounds, that are toxic to targeted crop pests but do not harm humans, animals, fish, birds or beneficial insects. Because biopesticides act in unique ways, they can control pest populations that have developed resistance to conventional pesticides.

A biopesticide farmers (including organic farmers) have used since the 1930s is the microorganism Bacillus thuringiensis, or Bt, which occurs naturally in soil. Several of the proteins the Bt bacterium produces are lethal to certain insects, such as the European corn borer, a prevalent pest that costs the United States $1.2 billion in crop damage each year. Bt bacteria used as a biopesticidal spray can eliminate target insects without relying on chemically based pesticides.

Using the flexibility provided by biotechnology, we can transplant the genetic information that makes the Bt bacterium lethal to certain insects (but not to humans, animals or other insects) into plants on which that insect feeds. The plant that once was a food source for the insect now kills it, lessening the need to spray crops with chemical pesticides to control infestations.

Herbicide Tolerance

Good planting conditions for crops will also sustain weeds that can reduce crop productivity as they compete for the same nutrients the desired plant needs. To prevent this, herbicides are sprayed over crops to eliminate the undesirable weeds. Often, herbicides must be applied several times during the growing cycle, at great expense to the farmer and possible harm to the environment.

Using biotechnology, it is possible to make crop plants tolerant of specific herbicides. When the herbicide is sprayed, it will kill the weeds but have no effect on the crop plants. This lets farmers reduce the number of times herbicides have to be applied and reduces the cost of producing crops and damage to the environment.

Increasing Yields

In addition to increasing crop productivity by using built-in protection against diseases, pests, environmental stresses and weeds to minimize losses, scientists use biotechnology to improve crop yields directly. Researchers at Japan's National Institute of Agrobiological Resources added maize photosynthesis genes to rice to increase its efficiency at converting sunlight to plant starch and increased yields by 30 percent. Other scientists are altering plant metabolism by blocking gene action in order to shunt nutrients to certain plant parts. Yields increase as starch accumulates in potato tubers and not leaves, or as oil-seed crops, such as canola, allocate most fatty acids to the seeds.

Biotechnology also allows scientists to develop crops that are better at accessing the micronutrients they need. Mexican scientists have genetically modified plants to secrete citric acid, a naturally occurring compound, from their roots. In response to the slight increase in acidity, minerals bound to soil particles, such as calcium, phosphorous and potassium, are released and made available to the plant. Nitrogen is the critical limiting element for plant growth and, step-by-step, researchers from many scientific disciplines are teasing apart the details of the symbiotic relationship that allows nitrogen-fixing bacteria to capture atmospheric nitrogen and provide it to the plants that harbor them in root nodules.

- Plant geneticists in Hungary and England have identified the plant gene and protein that enable the plant to establish a relationship with nitrogen-fixing bacteria in the surrounding soil.
- Microbial geneticists at the University of Queensland have identified the bacterial gene that stimulates root nodule formation.
- Collaboration among molecular biologists in the European Union, United States and Canada yielded the complete genome sequence of one of the nitrogen-fixing bacteria species.
- Protein chemists have documented the precise structure of the bacterial enzyme that converts atmospheric nitrogen into a form the plant can use.

Resistance to Environmental Stresses

In addition to the biological challenges to plant growth and development just described, crops plants must contend with abiotic stresses nature dispenses regularly: drought, cold, heat and soils that are too acidic or salty to support plant growth. While plant breeders have successfully incorporated genetic resistance to biotic stresses into many crop plants through crossbreeding, their success at creating crops resistant to abiotic stresses has been more limited, largely because few crops have close relatives with genes for resistance to these stresses. The crossbreeding limitation posed by reproductive compatibility does not impede crop biotechnology; genes found in any organism can be used to improve crop production. As a result, scientists are making great strides in developing crops that can tolerate difficult growing conditions. For example, researchers have genetically modified tomato and canola plants that tolerate salt levels 300 percent greater than nongenetically modified varieties. Other researchers have identified many genes involved in cold, heat and drought tolerance found naturally in some plants and bacteria. Scientists in Mexico have produced maize and papaya that are tolerant to the high levels of aluminum that significantly impede crop plant productivity in many developing countries.

Crop Biotechnology in Developing Countries

Today, 70 percent of the people on the planet grow what they eat, and, despite the remarkable successes of the Green Revolution in the 1960s, millions of them suffer from hunger and malnutrition. Continuing population growth, urbanization, poverty, inadequate food distribution systems and high food costs impede universal access to the higher yields provided by technological advances in agriculture. In addition, the crops genetically improved by plant breeders who enabled the Green Revolution were large-volume commodity crops, not crops grown solely by small-scale subsistence farmers. For many farmers in developing countries, especially those in sub-Saharan Africa, the Green Revolution never materialized because its agricultural practices required upfront investments— irrigation systems, machinery, fuel, chemical fertilizers and pesticides—beyond the financial reach of small-scale farmers.

Today's biological agricultural revolution is knowledge intensive, not capital intensive, because its technological advances are incorporated into the crop seed. As a result, small-scale farmers with limited resources should benefit. In addition, because of the remarkable flexibility provided by crop biotechnology, crop improvement through genetic modification need no longer be restricted to the large-volume commodity crops that provide a return on industrial R&D investments. A beneficial gene that is incorporated into maize or rice can also be provided to crops grown by subsistence farmers in developing countries because the requirement for plant reproductive compatibility can be circumvented.

Realizing biotechnology's extraordinary capacity for improving the health, economies and living conditions of people in developing countries, many universities, research institutions, government agencies and companies in the industrialized world have developed relationships for transferring various biotechnologies to developing countries. The nature of the relationships varies, depending on the needs and resources of the partners involved. For example:

- Cornell University donated transgenic technology for controlling the papaya ring spot virus to research institutions in Brazil, Thailand and Venezuela and provided their scientists with training in transgenic techniques.

- Japan's International Cooperation Agency built tissue culture facilities at an Indonesian research institution so that scientists there could develop disease-free potato materials for planting. The Indonesian researchers are also working with scientists at Michigan State University to develop insect-resistant potatoes and sweet potatoes.
- An Australian agricultural research center collaborated with Indonesian researchers on studies of nitrogen fixation and development of disease-resistant peanuts.
- Seiberdorf Laboratories (Austria) worked with the Kenyan Agricultural Research Institute to transfer technology for cassava mutagenesis and breeding.
- Monsanto has donated virus resistance technologies to Kenya for sweet potatoes, Mexico for potatoes and Southeast Asia for papaya and technology for pro-vitamin A production in oilseed crops to India.
- Pioneer Hi-Bred and the Egyptian Agricultural Genetic Engineering Research Institute (AGERI) collaborated to discover potentially novel strains of Bt in Egypt. Pioneer trained AGERI scientists in methods for characterizing Bt strains and transgenic techniques. Patents are owned by AGERI and licensed to Pioneer.
- AstraZeneca trained scientists from Indonesia's Central Research Institute for Food Crops in the use of proprietary technologies for creating insect-resistant maize.
- The Malaysian palm oil research institute has collaborated with Unilever and universities in England, the United States and the Netherlands on research to change the nutritional value of palm oil and find new uses for it, such as lubricants, fuels, a vitamin E precursor, natural polyester and biodegradable plastics.

While technology transfer has been and, no doubt, will continue to be an essential mechanism for sharing the benefits of crop biotechnology, many developing countries are taking the next step: investing resources to build their own capacity for biotechnology research, development and commercialization. The leaders in these countries recognize the potential of crop biotechnology to provide agricultural self-sufficiency, preserve their natural resources, lower food prices for consumers and provide income to their small farmers. Even more important, they understand that biotechnology has the potential to improve existing exports and create new ones, leading to a more diversified economy and increased independence.

But they also know that many of their agricultural problems are unique and can best be solved by local scientists who are familiar with the intricacies of the problems, local traditions, and applicability—or lack of it—of technologies that were developed to solve agricultural problems in industrialized countries. To move their countries forward, they are investing human and financial resources in developing local strength in crop biotechnology. For example:

- The Malacca government in Malaysia formed a unit in the Chief Minister's Office to promote research and development in biotechnology and established the Sarawak Biodiversity Center to ensure sustainable use of genetic resourcesand to build a strong database for bioresources.
- Taiwan opened an extension of the Hsinchu industrial park devoted exclusively to biotechnology. Companies in the park will have access to $850 million in government research and development funds and $4 billion in state and private venture capital, plus a wide range of support services including marketing and global patent applications.
- Pakistan's Ministry of Science and Technology prepared a biotechnology action plan and funded a three-year program to promote biotechnology research and development.
- Uganda's National Council of Science and Technology established its first commercial agricultural biotechnology lab to produce disease-free coffee and banana plantlets.
- Egypt's government, a longtime supporter of agricultural biotechnology, released a report encouraging farmers to plant genetically modified crops to benefit from reduced pesticide applications, lower production costs, higher yields and increased income.

Environmental and Economic Benefits

Beyond agricultural benefits, products of crop biotechnology offer many environmental and economic benefits. As described above, biotech crops allow us to increase crop yields by providing natural mechanisms of pest control in place of chemical pesticides. These increased yields can occur without clearing additional land, which is especially important in developing countries. In addition, because biotechnology provides pest-specific control, beneficial insects that assist in pest control will not be affected, facilitating the use of integrated pest management. Herbicide-tolerant crops decrease soil erosion by permitting farmers to use conservation tillage.

Because farmers in many countries have grown biotech crops for years, data is now available for assessing the magnitude of the environmental and economic benefits provided by biotechnology. In the past few years, a number of independent researchers have produced reports documenting these benefits.

According to the National Center for Food and Agricultural Policy's (NCFAP) 2004 report, in 2003 the 11 biotech crop varieties adopted by U.S. growers increased crop yields by 5.3 billion pounds, saved growers $1.5 billion by lowering production costs, and reduced pesticide use by 46.4 million pounds. Based on increased yields and reduced production costs, growers realized a net economic impact or savings of $1.9 billion. Three new traits for corn and cotton were introduced in 2003, and the NCFAP study takes into account six biotech crops—canola, corn, cotton, papaya, soybean and squash.

In its report "Conservation Tillage and Plant Biotechnology," the Conservation Tillage Information Center (CTIC) at Purdue University attributes the recent improvements in tillage reduction to the increased use of the herbicide-tolerant varieties produced through biotechnology. CTIC concludes that the increase in conservation tillage associated with herbicide-tolerant crops decreases soil erosion by 1 billion tons of soil material per year, saves $3.5 billion per year in sedimentations costs and decreases fuel use by 3.9 gallons per acre.

According to the International Service for the Acquisition of Agri-Biotech Applications, a single biotech crop, Bt cotton, has led to the following environmental and economic benefits for farmers in developing countries:

- From 1999 to 2000 in China, insecticide usage decreased by 67 percent and yields increased by 10 percent, leading to income gains of $500 per hectare.
- Extensive field trials in India from 1998 to 2001 demonstrated a 50 percent reduction in insecticide spraying and a 40 percent increase in yields, which equals an increase in income from $75 to $200 per hectare.

Small farmers in South Africa gained through a 25 percent yield increase and decreased number of insecticide sprays from 11 to four, reducing pesticide costs by $45 per acre. The higher cost of Bt seed (up to $15 per hectare for small farmers) resulted in an average economic advantage of $35 per hectare.

Regulation of Crop Biotechnology

Since combining specific genes from donor and host plants does not alter the basic nature of the host plant, the result of genetic modification is predictable and can be carefully controlled. As with any new variety of food, the developers test extensively for safety, quality and other factors.

U.S. regulatory policy for biotechnology products was established in 1986 with the publication by the White House Office of Science and Technology Policy of the "Coordinated Framework." This framework builds on the work of international expert bodies (such as the Organization for Economic Cooperation and Development [OECD] and the U.S.

National Academy of Sciences). The responsibilities of regulatory agencies are clarified, linked to the laws they administer and coordinated with other agencies that have potentially overlapping responsibilities.

The U.S. Food and Drug Administration (FDA) approves the safety of all foods and new food ingredients. In addition, all producers are required to ensure the safety and quality of anything they introduce into the food supply.

The FDA requires strict premarket testing and regulatory oversight of genetic modifications that significantly alter the nutritional value of the host food, use genetic material from outside the traditional food supply or use known allergens.

The FDA also requires labeling of any food product produced through biotechnology that significantly alters the host food's nutritional value or uses material from a known allergen. For example, any product that uses a gene from a peanut, which is a potential allergen, would be subject to testing and labeling requirements. The FDA also has the authority to order unsafe products off the market.

The USDA and the U.S. Environmental Protection Agency (EPA) impose safety requirements and/or performance standards on the development of pesticides, herbicides and genetically enhanced test crops. The USDA regulates to ensure that crop varieties improved through biotechnology are at least as safe as those produced through traditional breeding programs. Rigorous assessments are conducted concerning the derivation of the new varieties and their performance under contained and controlled field trials. Each field trial is subjected to review for site-specific safeguards. This extensive regulatory oversight has resulted in more than 5,000 field trials and commercial plantings on over 20,000 test and commercial plots in the United States alone over the past 15 years.

The EPA also coordinates with the USDA and FDA, using its own statutes to regulate the growing of plants with pestprotection characteristics. The EPA sets allowable food residue tolerance levels for any novel compounds that might be used.

Forest Biotechnology

Throughout the world, wood provides us with fuel, construction materials and paper, and its supplies are dwindling rapidly. Wood products are currently a $400 billion global industry, employing 3 million people. Demand for wood products is expected to increase, even as major economies, such as Europe and Japan, are unable to grow enough trees to meet their current demand. According to the U.N. Food and Agriculture Organization, world demand for wood products in 2010 will be about 1.9 billion cubic meters, almost 20 percent higher than it is now. We must attempt to meet that demand without cutting down the world's remaining forests.

Increasing Productivity

We are using biotechnology to create disease and insect-resistant trees and to increase their growth rates. Scientists are also learning how to use biotechnology to improve the efficiency with which trees convert solar energy into plant material and to shunt more of that energy into wood production and less into pollen, flowers or seeds. All of these methods of increasing productivity should decrease the pressure on natural forests.

However, developing trees through the use of biotechnology is a lengthy undertaking because trees take a long time to grow. So, researchers are looking to other methods for increasing productivity. For example, they are using a biotechnology process in a fungus to fight diseases that infect trees and are working on improving the microorganisms that live on tree roots and provide trees with nutrients, much as nitrogen-fixing bacteria increase the nutrients available to soybeans and alfalfa. In addition, biopesticides have also been used extensively to control forest pests, and we expect progress in insect cell culture to boost the number of biocontrol agents available for forest insect control.

Environmental Benefits

Perhaps a more important economic role for biotechnology in this industry will be found in its changing the way we convert trees to useful products. Extensive research is being conducted to increase a tree's amount of cellulose, the raw material for papermaking, and to decrease the amount of lignin, a tough molecule that must be removed in papermaking.

Traditionally, removing lignin from trees has required harsh chemicals and high energy costs, so changing the cellulose: lignin ratio genetically has important environmental implications, as does increasing the growth rate of tress. Because trees absorb carbon dioxide, any advance that allows us to increase tree yields without cutting down forest could have significant positive effects on global warming. Other environmental benefits that biotechnology is providing to the forestry industry include enzymes for

- pretreating and softening wood chips prior to pulping.
- removing pine pitch from pulp to improve the efficiency of paper-making.
- enzymatically bleaching pulp rather than using chlorine.
- de-inking of recycled paper.
- using wood-processing wastes for energy production and as raw materials for manufacturing high-value organic compounds.
- remediating soils contaminated with wood preservatives and coal tar.

Animal Biotechnology

Animals are playing a growing role in the advancement of biotechnology, as well as increasingly benefiting from biotechnology. Combining animals and biotechnology results in advances in four primary areas:

1. Improved animal health through biotechnology.
2. Advances in human health through biotechnology studies of animals.
3. Enhancements to animal products with biotechnology.
4. Environmental and conservation efforts of biotechnology.

Animal biotechnology includes all animals—livestock, poultry, fish, insects, companion animals and laboratory animals—and applications of the scientific tools of genomics, transgenics, and cloning technologies.

How Are Products of Animal Biotechnology Regulated?

Three government agencies regulate the animal health industry: the U.S. Department of Agriculture regulates veterinary biologics, vaccines and diagnostic test kits; the Food and Drug Administration reviews and approves new pharmaceuticals and feed additives; and the Environmental Protection Agency regulates pesticides and topical products that kill fleas and other parasites.

The regulatory processes for the products of animal biotechnology are being reviewed by the Office of Science and Technology Policy, seeking coordination among the federal agencies for a science-based, streamlined approach. Little published regulatory guidance exists for many of the products being developed.

In 2003, the U.S. Food and Drug Administration's Center for Veterinary Medicine published a draft risk assessment regarding cloning of farm livestock and the safety of cloned-animal food products for human consumption. The FDA concluded that meat and milk from animal clones were safe to eat. Next steps include finalizing the risk assessment (expected in 2005) and proposing a risk management process.

Using Biotechnology to Improve Animal Health

The market for biotechnologybased animal health products and services is estimated to be $2.8 billion and expected to grow to $5.1 billion in 2005. As of July 2003, there were 111 animal biotech products, including bacterins and killed virus vaccines. The animal health industry invests more than $400 million a year in research and development.

Farm Animals: Livestock and Poultry

Biotechnology provides new tools for improving animal health and increasing livestock and poultry productivity. These improvements come from the enhanced ability to detect, treat and prevent diseases and other problems; from better feed derived from transgenic crops designed to meet the dietary needs of different farm animals; and from improved animal breeding.

The animal health industry has developed many effective treatments that can prevent and treat dangerous diseases that could potentially strike entire livestock herds and poultry flocks. Quick diagnosis and treatment, coupled with strong preventative measures, help lower production costs and improve overall animal well-being. Additionally, healthier farm animals result in safer foods for consumers.

- Biotechnology allows farmers to quickly diagnose the following infectious diseases through DNA and antibody-based tests: brucellosis, pseudorabis, scours, foot and-mouth disease, bluetongue, avian leucosis, mad cow disease and trichinosis.
- Farmers may soon be able to manage several farm animal diseases through biotechnology based pharmaceuticals, including foot-and-mouth disease, classical swine fever and bovine spongi form encephalopathy.
- New biological vaccines protect farm animals from a wider range of diseases, including foot-and-mouth disease, scours, brucellosis, shipping fever, lung infections affecting pigs (pleuro-pneumonia, pneumonic pasteurellosis, enzootic pneumonia), hemorrhagic septicemia, fowl cholera, Newcastle disease of poultry, rabies, and infections that affect cultivated fish.
- In addition to these existing vaccines, work is being doneto develop a vaccine for an African cattle disease called East Coast fever. If successful, this vaccine would be the first against a protozoan parasite and could lead to the development of a malaria vaccine for humans.
- Molecular-based typing of pathogens, such as genetic fingerprinting, allows for the monitoring of the spread of disease within and between herds and can identify the source of an outbreak.
- Genetic analysis of animal pathogens is leading to an improved understanding of the factors that cause disease and how best to control them.
- Crops improved through biotechnology provide nutritionally enhanced feed for farm animals, with the addition of amino acids and hormones, to improve animal size, productivity and growth rates. Through biotechnology, many of these feeds can increase digestibility of low quality roughage. Scientists are working on new crops to develop feed with edible vaccines for farm animals. In the near future, pigs could be fed transgenic alfalfa that would stimulate immunity to a serious intestinal virus

- Researchers are developing a vaccine alternative to castration for livestock. Bull calves are castrated to control aggression, and male pigs are castrated to avoid "boar taint," which makes their meat inedible. The new vaccine will render the animals sterile and eliminate the need for surgery, while ensuring the animals grow well.

In addition to diagnostic tests, vaccines and medicines for farm animals, biotechnology plays a growing role in farm animal breeding programs. With genetic mapping techniques, genetically disease-resistant animals can be identified and used for breeding programs, resulting in naturally healthier offspring. Conversely, animals with some of the following genetic weaknesses and defective genes can be identified and removed from breeding programs:

- New DNA tests can identify pigs with the genetic condition porcine stress syndrome, which causes tremors and death under stressful conditions.
- Inherited weaknesses of cattle can be identified with DNA tests, which are currently being used in national breeding herds in Japan. Tests can identify leukocyte adhesion deficiency, which causes repeated bacterial infections, stunted growth and death within the first year of life. Factor 13 deficiency, which prevents blood from coagulating normally, can also be identified. Other DNA tests can identify a hereditary condition that produces anemia and retarded growth in Japanese black cattle.

Increasing Livestock Productivity

Livestock producers are always interested in improving the productivity of agricultural animals. Their goal is to obtain the same output (milk, eggs, meat, wool) with less input (food), or increased output with the same input. Increasing muscle mass and decreasing fat in cattle and pigs have long been goals of livestock breeders.

Using biotechnology to increase the productivity of livestock is a variation of selective breeding. We select individual animals that possess desirable traits; then, instead of breeding the animals, we collect eggs and sperm and allow fertilization to occur in a laboratory dish. This in vitro fertilization is followed by embryo culture, a form of mammalian cell culture in which the fertilized egg develops into an embryo. When the embryo is a few days old, it is taken from the laboratory dish and implanted into a female of the same species—but not necessarily of the same breed. This is known as embryo transplant.

Sometimes, the embryo, which is a clump of cells at this stage in development, is divided into several parts, and each cell cluster is implanted. This is a form of cloning that has been used for a few decades to improve the genetic makeup of the herd more quickly than by simply relying on a single female that produces one calf per year.

Genomics technology is being applied to improving the conventional breeding of superior animals in order to produce desirable traits. In 2003, the first validated SNP beef cattle genome was created. SNP technology is being used to identify clusters of genes that contribute to a trait—for example, leaner beef cattle. Then, through conventional breeding, lines of cattle are being developed that express the increased muscling. Worldwide, research teams are working to sequence the genomes of a wide variety of animals and insects, including mice, dogs, sheep, swine, rats and fruitflies. In October 2004, the Bovine Genome Sequencing Project announced it had successfully sequenced the cow genome. In December 2004, the Chicken Genome Sequencing Consortium announced it had sequenced the chicken genome.

Additional Applications of Animal Agriculture

The biotechnology industry has proposed solutions to food safety problems such as emerging animal diseases and food-borne pathogens. DNA sequencing of individual animals could serve as the ultimate animal identification, allowing for tracking of meat from farm to table. Experimental cattle resistant to bovine spongiform encephalopathy are being produced using biotechnology techniques such as knock-out technology and cloning. A cattle vaccine produced in plants could reduce Escherichia coli 0157:H7 shedding in feedlot cattle, a further assist toward improved food safety on the farm.

Also:

- Genetic mapping and the development of DNA markers are being used to identify genes in chickens that have developed a resistance to Marek's disease, a virusinduced disease similar to cancer.
- Scientists have discovered a gene in certain sheep that converts food into muscle 30 percent more efficiently.
- Researchers are developing biotech plants that produce vaccines for hoof-and-mouth disease, swine gastroenteritis and rabbit hemorrhagic disease.
- Recombinant protein vaccines are under development to control coccidiosis in chickens, which seriously impairs their growth.
- DNA cloning technology is being used to research treatments for cryptosporidiosis, a parasitic disease found mostly in calves, but also in children and immunosuppressed individuals, such as those with AIDS.
- Sodium hyaluronate products, used in the treatment of joint disease in horses and other farm animals, decrease the effect of inflammatory enzymes in cartilage. In addition, a patented biofermentation process is now available that allows these therapies to be given intravenously.

- Many animal feed products are enhanced with proteins to boost nutrition and control disease.
- Injectable products are being used to protect cattle from 36 stages of internal and external parasites, including ostertagi, nematodes and trematodes.

Companion Animals

Approximately 137 million dogs and cats are companion animals in the United States (in more than 60 percent of all American households). America's emotional attachment to its pets is evidenced by the estimated $29.5 billion spent on U.S. pets in 2002. To help companion animals live longer and healthier lives, the greatest expense for pet owners are services related to veterinary care and healthcare products, which accounted for more than $7 billion in 2000.

The animal health industry has developed products that contribute to the well-being of companion animals. Pets benefit from preventive medicines and disease treatments that have been improved through biotechnology. Animal vaccines developed through biotechnology are critical to preventing diseases such as rabies, distemper, feline leukemia and hepatitis. In addition, researchers have developed biotechnology-based products to treat heartworm, arthritis, parasites, allergies, dental problems, heart disease, kidney failure, separation anxiety, cognitive dysfunction syndrome and other problems.

Recent biotechnology-driven developments in companion animal health care include the following:

- Immunologists have developed a vaccine for feline immunodeficiency virus (FIV), an organism carried by as many as 25 percent of cats. In addition to saving cats' lives, the research for creating the FIV vaccine provides many clues in the development of an HIV/AIDS vaccine.
- Gene therapy has restored the vision of dogs afflicted by Leber congenital amaurosis, an untreatable condition that causes almost complete blindness. Currently researchers are successfully testing gene therapy for melanoma, canine lymphoma and bone cancer. Canine cancer lymphoma accounts for 20 percent of all canine tumors and kills most dogs within a month of diagnosis. Gene therapy treatment may prolong life by a year.
- A rabies vaccine has been widely used with wild raccoon populations to limit transmission to companion animals. In the United States, an estimated 40,000 people undergo treatment for rabies annually at an average cost of $1,650.
- Projects for mapping the genetic code of fleas may someday result in products that rid dogs and cats of the insect.

- Monoclonal antibody technology is being used to develop treatments for canine lymphoma.
- Other antibody technology is being used to develop diagnostics for feline infectious peritonitis and feline immunodeficiency virus.

Other recent biotechnologydriven developments in companion animals are listed below:

- An allergen-free cat is being developed.
- The first biotech animal to be sold to the public reached the market in January 2004; GloFish are biotech ornamental fish that contain a gene from a sea anemone. Under black light, the GloFish fluoresce in a brilliant red color. The U.S. Food and Drug Administration conducted a complete scientific and technical review of the biotech fish, including assessing target animal safety, human safety and environmental safety, and found them safe and environmentally harmless.

Animal Biotechnology to Enhance Human Medical Applications

Animals are often used as models for research as many of the technologies developed for animals can be transferred to humans. Some of the work being done with animals that will advance human health:

Xenotransplantation

Extensive research has been done on the potential for using biotech animals as blood or organ donors for humans. The primary barriers to successful xenotransplantation include the immune reactions of the recipient from the graft, the possibility that animal tissues or organs might not function well in a human recipient, and the possibility that the xenotrans plant might convey infections. Biotechnology has been used to address the problem of immunorejection, and biotech pigs have been developed with organs that may resist rapid rejection by the human immune system.

"Pharm" Animals

Researchers are developing biotech animals, including cows, goats and sheep, that produce milk containing therapeutic proteins. These proteins may be used to nourish premature infants or to treat emphysema, cystic fibrosis, burns, gastrointestinal infections and immunodeficiency diseases such as AIDS. Some interesting ongoing projects include:

- Biotech goats that produce milk containing tissue plasminogen activators (TPA), which can dissolve clots in heart attack victims.
- Dutch researchers are working with biotech rabbits that secrete a potential

drug for Pompe's disease in their milk. Pompe's disease is an extremely rare genetic disorder that can result in crippled muscles, breathing problems and sometimes death.

- Scientists are working with biotech goats that produce an experimental anticancer medication.
- Biotech cows can now produce the human milk protein lactoferrin, which is an antibacterial protein that can be used to treat immunosupressed patients or be incorporated into infant formula.

Genetic Sequencing Projects

Numerous genetic sequencing projects are being conducted; understanding the human genome better will lead to the development of new ways to treat disease.

Enhancing Animal Products

Biotechnology can make dramatic improvements to animal products that humans consume and use. Some of these improvements result from vaccines, medicines and diagnostic tests that make animals healthier. However, biotechnology has also made great strides in enhancing animal products at a cellular level through transgenic and cloning technology. Some of these enhancements include:

- Researchers can produce biotech cows, pigs and lamb with reduced fat and increased lean muscle.
- Genetic mapping projects allow farmers to identify highly productive animals for breeding programs.
- Vaccines have found new uses and can now improve egg production in breeding turkeys. The vaccine stimulates the turkey's immune system to overcome the tendency to stop laying eggs.
- Other vaccines can improve the efficiency of feed conversion or modify hormone production to increase growth rates. Some vaccines can stimulate milk production or produce leaner meat.
- Biotech cows can now produce "designer milks" with increased levels of protein that can improve the diet of children or affect production of cheese and yogurt. Additionally, scientists are now working to remove from milk the proteins that cause lactose intolerance. It is estimated that 90 percent of the Asian population is lactose intolerant.
- Australian scientists have increased wool production by feeding sheep biotech lupin, a mainstay of sheep's summer diet.
- Scientists are working to develop biotech shrimp that lack the protein responsible for 80 percent of shrimp allergies.

Studies conducted by the National Academy of Sciences (NAS) have determined that cloned and transgenic animals and their products are safe for human consumption.

Biotech versions of several animal-feed crops are under study. These products are designed to improve the quality of protein, oils or energy availability in the final animal food product. One crop is designed to improve shelf life of beef by improving the antioxidant properties of the meat's fats.

Environmental and Conservation Efforts

Environmental Impacts

Livestock producers are challenged with identifying how to dispose of more than 160 million metric tons of manure annually. Animal manure, especially that of swine and poultry, is high in nitrogen and phosphorus, which can contribute to surface and groundwater pollution. Several crops improved with biotechnology may offer animal feed that decreases phosphorus and nitrogen excretion, total manure excretion and offensive odors.

Further, the Enviro—Pig is a biotech pig that has a gene added to enhance salivary phytase, thereby improving phosphorus digestibility and retention of phosphorus in pork, with reduced excretion of phosphorus in the manure of the animal. The goal is to reduce the chance of manure contributing to groundwater contamination in areas that surround livestock farms.

Endangered Species Conservation

Biotechnology is also providing new approaches for saving endangered species.

Reproductive and cloning technologies, as well as medicines and vaccines developed for use in livestock and poultry, can also help save endangered mammals and birds.

Borrowing biotechnology techniques used by livestock breeders, veterinarians at the Omaha zoo recently used hormonal injections, artificial insemination, embryo culture and embryo transfer to produce three Bengal tiger cubs. A Siberian tigress served as the surrogate mother for these embryos.

Worldwide, researchers have used cloning technologies to conserve endangered species. In September 2001, researchers at the University of Teramo, Italy created the first viable clone of an endangered species, the European mouflon. There are thought to be fewer than 1,000 adult mouflons in Sardinia, Corsica and Cyprus, and it is the smallest wild sheep in the world. In January 2001, the world's first cloned endangered species, an ox-like guar, was born in the United States, though it succumbed to a common dysentery infection. There are estimated to be less than 36,000 guars in India and Southeast Asia due to human development of their natural habitat. Researchers have also worked to clone the argali, the largest wild sheep, but have been unable to produce live offspring. In December 2003, the first cloned whitetail deer was reported in the United States. Though not an endangered species, researchers believe the successful clone will provide valuable insight into cloning other wild animals, including endangered species.

In April 2003, the San Diego Zoo reported the birth of a cloned banteng, a wild cow native to the island of Java. Since January 2004, the banteng has been viewable to the public at the San Diego Zoo; it is the first cloned species to be on display to the public at any zoo. Researchers at the San Diego Zoo also employ other biotech and reproductive technologies in their conservation efforts. In 1975, they created the "Frozen Zoo," a genetic bank that currently houses frozen cells from over 7,000 endangered or threatened mammals, birds and reptiles. Other animal conservation organizations, including the Zoological Society of London and the Cincinnati Zoo, have created genetic databases to store cryogenically frozen samples of DNA, gametes and cell tissues for later use.

Recently, Chinese scientists announced that they are close to cloning the Giant Panda using trans-species cloning technology. The Giant Panda is a highly endangered species.

Biotechnology techniques for working with endangered species have not been limited to cloning. Some researchers are using genetic samples to study the distribution of species and track the relationships between different groups of animals. These studies may help to prevent excessive interbreeding among small groups of animals.

Genetic studies can also help produce a healthier population of endangered species through increased genetic diversity. Conservationists studying the endangered Florida panther realized that, as the population shrank, inbreeding became more common. Through genetic testing, researchers found that the panthers were closely related to Texas cougars and had previously interbred. By introducing some cougars in the Florida panther breeding pool, scientists increased the genetic diversity of the species, resulting in a healthier panther population.

Endangered Plants

Endangered plants may also benefit from the flexibility in problem-solving biotechnology provides. Scientists are developing strategies for resurrecting the American chestnut tree, brought to virtual extinction by chestnut blight, and restoring the Cornish elm tree, 90 percent of which have been destroyed by Dutch elm disease, to Great Britain.

One approach to regenerating populations of chestnuts involves using genomics to identify and isolate genes for blight resistance found in the Chinese chestnut, then adding those genes to the chestnut seedlings that continue to sprout from the trunks of dead chestnut trees. The time frame for creating blight-resistant American chestnuts shrinks to less than five years if plant cell culture and recombinant DNA technology are used.

In the U.K. scientists are considering a number of biotechnology options for restoring elms: identify viruses that infect the fungus that causes Dutch elm disease and use them as biocontrol agents; genetically alter the tree with the Bt gene that specifically kills beetles; and genetically alter the tree to produce substances that will kill the fungus.

Aquaculture

Aquaculture is the growth of aquatic organisms in a controlled environment. The increased public demand for seafood, combined with the relatively small supply of aquaculture products provided by U.S. companies, has encouraged scientists and industry to study ways that marine biotechnology can increase the production of marine food products. By using biotechnology techniques, including molecular and recombinant technology, aquaculture scientists study the growth and development of fish and other aquatic organisms to understand the biological basis of traits such as growth rate, disease resistance or resistance to destructive environmental conditions.

Researchers are using marine biotechnology to identify and combine valuable traits in parental fish and shellfish to increase productivity and improve product quality. The traits scientists and companies are investigating for possible incorporation into several marine organisms include increased production of natural fish growth factors and the natural defense compounds marine organisms use to fight microbial infections. Biotechnology is also improving productivity through the development of feed additives, vaccines and other pharmaceutical agents.

Biotech Salmon

Salmon enhanced through biotechnology have received much public exposure over the last year from reports issued by the National Academy of Sciences (NAS) and the Pew Initiative on Food and Biotechnology. Additionally, in 2003, four states passed legislation banning the production of biotech fish in their waterways, including Maryland, California, Oregon and Washington. Several state legislatures have strengthened or are reviewing regulations for non-native and biotech fish, including Michigan and Florida. Some of the biotech improvements being made with fish include:

- Some biotech salmon reach maturity quickly and do not hibernate, which enables year-round availability of salmon.
- Researchers are trying to develop fish that are more resistant to disease, tolerant of low oxygen levels in the water and tolerant to freezing temperatures.
- Some species of fish naturally produce a protein that allows them to survive

in the Arctic. This "anti-freeze" gene has been transplanted to other species of fish so they can survive in very cold waters.

Biotecnological Tools for Increasing Productivity of Pulse Crops

Pluses play a significant role in Indian agriculture because they provide protein rich diet to poor people of the country. They contain 20-29% protein i.e. about 2.5 times more than cereals. Inspite of release of large number of high yielding varieties, extension of irrigated areas and use of agrochemical, growth recorded in total production of pulses from 11.82 million tonnes in 1970-71 to 13.19 million tonnes in 1995-96 is considered to be non revolutionary. This resulted in decline in per capita availability of pulses to 37g/person/ day. There are several reasons for stagnation in pulses production such as narrow genetic base, lack of genetic variability in useable genepool, inability to transfer increased input to increase output (seed yield) susceptibility to large number of biotic and biotic stresses, photoperiod sensitive and non synchronous flowering and maturity. It has been estimated that on account of insect pest complex infesting pulse crops alone, nearly 2.0 to 2.5 million tonnes of pulses is lost annually. The monitory value of this loss will be approximately Rs.3000 to 3750 crores.

Conventional methods of crop improvement have not improve productivity of pulses with significant pace due to serveral reasons. These methods are slow, expensive and time consuming. Besides lack of reliable resistant sources, undersirable linkages and long gestation period are considered to be the main limitation of these methods. Hence, there is dire need to introduce innovative biotechnological approaches to enhance pulse productivity to meet out the market demand.

Biotechnology and genetic engineering has emerged as potential tool in recent year for genetic manipulation of crop plants. The said technology promises to improve crop productivity through complementing traditional breeding by decreasing dependence on harmful chemicals pesticides. Fertilizers and antibiotics etc. The tools of biotechnology are now available to plant breeders and can be used effectively to supplement conventional practices in improvement of pulse crops. It is expected that products of biotechnology will be ecologically sage, economically feasible and technologically expectable to Indian farmers.

Efficient transportation and distribution Synthetic seeds of material. Effort have been successful to develop several intergeneric and interspecific hybrids using embryo/ ovule/ovary/endosperm culture in grain legumes, Several useful genes imparting resistance to various biotic and biotic stresses have been successfully introgressed on wild relatives of lathyrus, lentil, chickpea, pigeonpea urd bean and mung bean. The resistance to Aschochyta blight to chickpea has been successfully transformed from wild species Cicer judaicum to elite chickpea genotype PDG 84-10 at IIPR, Kanpur.

The techniques of in vitro mutagenesis in vitro selection and somaclonal variation are being also successfully employed to various pulse crops to develop promising genotypes resistant to various biotic and a biotic stresses. This technique has resulted success in developing. Aschochyta blight resistance in chickpea. Further early results in in vitro selection for salinity tolerance in chickpea is also very encouraging at IIPR, Kanpur. Similarly, the technology of cell suspension culture followed by selection of mutants has also been used successfully in major pulse crops at different centres in the country.

Identification and characterization of pathogens and pests is problematic and always pose challenge to plant pathologist and breeders in varietal development of pulse crops. The techniques of molecular diagnosis and DNA finger printing has facilitated resulted easy diagnosis and the precise characterization of pathogene isolates and insect biotypes at molecular level. This has increased the efficiency of breeding programmes. Important pathogenic like Aschochyta rabi and Fusarium oxisporum of chickpea have already been characterized using DNA fingerprinting techniques.

In a new world order and changed scenario the characterization of germplasm and plant material is very essential to protect out natural wealth. The techniques of DNA fingerprinting have been proved very handy in protecting the intellectual property rights. These techniques are now being successfully used in pulse crops to characterize and catalogue the accessions of gene bank to protect the genetic wealth.

DNA marker techniques are the new innovations and have tremendous potential in increasing the efficiency of conventional breeding programmes. DNA marker technique can be used for tagging important genes which can subsequently be used for early, easy and precise selection in segregating of promising plants populations. DNA markers have been found linked to wilt resistance in chickpea, powdery mildew resistance in peas and nemarode resistance in mung bean and urd bean.

Transgenic technology which involves insertion of foreign DNA sequence has tremendous potential for improvement of crop plants. Efforts are on to develop transgenic plants resistance to pod borer in chickpea and pigeroxpea using Bt. Crystal protein gene from a soil bacterium. Early results obtained at various centres in the country are very encouraging. Similarly, efforts are also being made to develop transgenic urdbean and mung bean using coat protein gene.

Current status and future strategies for genetic manipulation of pulse crops using biotechnological tools will be discussed in detail.

Biotechnology for the Cotton Farmer Problems and Prospects

Cotton or white gold as it is aptly called is grown for its lint and seed which yield cotton fibre and seed oil, respectively. This crop occupies 32-33 mha of world area with a production of 20-25 metric tonnes. In India its area spans over 7.5 mha with an average yield of 290 kg/ha of lint and 870 kg/ha of seed cotton. To meet the challenges of 2000AD with a population of more that 960 million, a total producing of 19 million bales is required as against the 13-14 million bales of today. This can be achieved by the use of improved crop production practices coupled with appropriate pest management tactics. In addition, generation of novel. Transgenics may help achieve the near impossible. Genes that have been identified as potentially profitable, if engineered into acceptable cultivars can be used to generate such transgenics. Among these are genes imparting resistance to herbicides, insects, pathogens and abiotic stresses. It is also widely accepted now that a number of other qualitative characters can be improved, such as fibre strength, fineness, colour and thermal adaptability of the fibre. Thansgenic plants have become realistic components of stress management world over. Bollworm and herbicide resistant transgenic cotton have received the approval of the Environmental Protection Agency and have been commercially released in the US for cultivation. Considering the fact that numerous biotic and abiotic stresses limit cotton production, it is likely that future strategies might orient towards the development of a multiadversity resistant high yielding transgenic cotton variety with superior fibre qualities. The most important aspects in the development of transgenic plants are:

- Identification, isolation and cloning of the desirable genes
- Transformation of plant cells or tissues with a suitable vector
- Regeneration of the transformed cells or tissue
- Confirmation of the gene integration and expression
- Hardening and field establishment of the transgenics

Major Achievements Worldover in the Field of Cotton Transgenics

1. Genes

Genes for Insect resistance used:

- Toxin genes from Bacillus thuringiensis
- Protease inhibitor genes from plants & the horn worm

Genes used for herbicide resistance:

- Entrails gene from a bacteria Klebsielia spp. for Bromoxynil resistance
- Mono-oxygenase gene from a bacteria Alcaligenes eutrophus for 2,4-D resistance
- Acstolactate synthase gene for resistance against sufonyl urea & imidazoline group
- ER_PSP synthase gene for glyphosate resistance.

Genes used to Improve fibre qualities:

- **Polyhydroxybutyrate & polyhydroxyacetate from bacteria**
- **Expression of indigo pigment in fibres**

Genes to be used:

- **Cholesterol oxidase genes against insects**
- **Heliothis stunt virus genes against Heliothis**
- **Genes encoding chitinases glucanases, attacins cecropins vital coat proteins etc. against diseases.**

2. Transformation

- **Agrobacterium medicated transformation**
- **Particle acceleration gene delivery systems have been successfully used all over the world.**

3. Regeneration

- **Callus fissue regeneration — mostly genotype specific**
- **Protoplasts were regenerated successfully**

Meristematic tissues were successfully regenerated

4. Expression

- **Enhanced expression was achieved using truncated forms of the full length genes, repeat copies of the propoter, modification of the bacterial codons to plant preferred type.**

5. Commercial Release

- **Bollworm resistant transgenics — NUCOTN 33M & NUCOTN 35B were released in the US and cotton transgenics — are in now under field cultivation in Australia, South Africa and China.**
- **8 lakh hectares were planted with the transgenic cottons in 1996 and 1997**
- **More than 8,000 hectares were damaged by bollworms.**
- **Transgenic cotton plants in Australia were also found to be attacked by the bollworms.**
- **Herbicide (Bromoxynil) resistant transgenics BXN 57 and BXN 53 were released in the US in 1996.**

6. Problems Encountered / Expected

- Gene silencing and expression instability
- Resistance in insects to the toxins used. Studies on baseline toxicity indicated that there is a natural variation in the ability of Helicoverpa armigera in tolerating the CrylAc toxins. Some strains are capable of surviving the totoxins by virtue of an in built capacity to tolerate the toxins, while others are not. While development of Helicoverpa resistance to the CrylAc toxins itself may not be of immediate concern, a slight shift in the tolerance levels resulting from continuous exposure to the transgenics, can cause control difficulties.
- High cost of transgenic seed. v* Microecological changes resulting in pest shifts. CrylAc is a broad spectrum lepidopteran toxins that can cause significant changes in almost all populations of lepidopteran insects occuring on the cotton crop. Some of these insects, which cause negligible economic damage to the crop, harbour parasitoids which have the potential to keep the bollworm populations under check. Consequently, in the absence or low populations of natural enemies, the bollworms can emerge as stronger pests that before. Insects pests such as Spodoptera litura which are less affected by the CrylAc toxins and which have been under check due to the use of pyrethroids can resurgace as major pests as pyrethroid use is likely to be reduced on transgenic Bt cottons.

7. Some Recent Development in Cotton Transgenics

- Three Bt transgenic lines, 95-1, RIOI and R-19 have been indigenously developed in China through Agrobacterium medicated (Somatic embryo) and pollen tube pathway transformation. Thestability of Bt gene expression and field efficacy have been confirmed.
- Microprojectile bombardment parameters for pollen mediated transformation and genetype independent protocols have been develop in China.
- A new methods of transformation through injection of Agrobacterium into developing embryos, has been reported from South Africa.
- Stable transformation and recrimination has been reported for Bt transgenics in Uzbekistan, China, Egypt and Australia. Pakistan reports development of cotton transgenic plants resistant torearcurl virus.
- An antisense DNA of CLCUV DNA-A bome ACI gone along with the antesense DNA of the AC2 and AC3 gene was used for the vector construction and transgenic cotton resistant to the CLCUV was developed in Pakistan.

- In Russia, A glucanase gene has been isolated from thermo philic bacteria and is being used as a new reporter system fortransgenic cotton development.
- Genetically engineered Baculovirus ACMNPV with genes from straw itch mite has been developed been developed by Monsanto and Zeneca with researchers from Madison, USA.
- Bt rransgenic cotton currently occupies 20% (85,000 ha) of cotton area in Australia. Isecticide use has been reduced by about 60% The efficacy has been moderate. The transgenic plants were found to express Bt proteimns only till the 95th day after which fluctuaions were observed. The reduction in expression was primarily due to down regulation which post transcriptional changes of the unstable RNA It was also reported that cotton tannins which increase with growth phase, act as antagonists with Bt toxins.
- After two years of field cultivation, true resistance to Bt has not yet been detected in Australia.
- Resistants to Bt in field populations of Helicoverpa armigera is reported form China
- Annually 100 million dollars are now being spent in the US, only in search of new insecticidal genes.
- In the US, 45% of cotton area this year will be under transgenic cotton (herbicide and insect resistant). However, the area under Bt cotton seems to on the decline.
- Transgenic cotton developed in Australia using the Stunt virus genes, is not performing well.
- Transgenic cotton developed in the US using lectin genes, is not performing well against Heliothis virscens.
- The Australian report that Helicoverpa armigera is at least 100 times less susceptible to Cry toxins as cinoared to Heliothis virescens.
- The Window strategy for insecticide resistance management in Australia is still important for transgenic cotton.
- Cotton Pest spectrum in the US is getting altered after the introduction of Bollgard. Unsprayed Bt cotton sustained 4 times more attack of tarnished bugs, 2.4 times more with boll weevil, 2.8 times more with stink bugs and Spodoptera.
- Due to these changes in pest complex farmers had to spray 3-5 times on boligard as compared to 6-8 times on non-Bt cottons.

Prospects and Potential

The application of biotechnonogy in cotton farming can be either in the form of production of fermentation products or novel recombinant products for use (example, recombinant Pseudomonas expressing CrylAc) or as transgenic plant with in buitt resistance to biotic and abiotic stresses. Transgenic crops with in built resistance to insect pests and diseases can be extremely useful as this would result in the reduction of insecticide use apart from making pest management simple for the farmer. Introduction of the bollworm resistant transgenic cotton is expected to reduce the use of chemicals used to control bollworm especially Helicoverpa importantly at a time when resistance to most insecticides including pyrethrolids is on the increase all over the world. Even if the introduction of the Bt cotton in India could result in a 25-30% reduction in insecticide use on cotton, this would mean a benefit of about of about Rs 300 crores, apart from the f-avourable impact on the environment so far, transgenic plants have been produced in about 60 plant species. Cotton has received special attention of the biotechnological companies in the developed countries who were attracted by the profit motives associated with the high value added to the transgenic seeds.

The firs Bt transgenic cotton has already been released, as Bollgard. The Delta and Pineland using DP 5410 and DP 5690 as recurrent parents. The D and PL brand BT transgenic were labelled as Nucotn 33B Two million acres were planted in 1996 in mid-south region of USA, mainly in regions where Heliothis virescens was problematic. Nearly 1.2 M acres were under Nucotn 33 B and the rest under Nucotn 35B.

The Bollgard brand Bt cotton seed was sold at US $34-36 as compared to US $8-9 per hectare for non-engineered cotton seed. The average cost of control of cotton insect pests in the US was approximately US $150 in the early 1990s. Hence the prices were still found to be attractive. The transgenics were found to be more effective against Helothis virescens as compared to pectinophora gossypiella and Heliothis zea.

Farmers growing transgenics had to sign an agreement with D & PL stating that they would not keep seed for planting next year. Quick ELISA tests have been devised to test for Bt toxins in plant parts, to check the illegal spread of transgenics.

Two more transgenic cotton varieties tolerant to herbicide Bromoxynil, trademarked as BXN 57 and BXN 53 were developed from the recurrent paren Coker 315 BYTHE Stonevuille pedigree company in collaboration with Calgene Inc. This seed was priced nearly 1.7 times the straight varieties.

In Australia the CSIRO and Narrabri research centre, in collaboration with Monsanto have developed transgenic cotton plants that were commercial released in 1997. This introduction resulted in 50-60 % reduction in the US $ 93 million spent by farmers each year on insecticides. Pakistan which uses 90% of its total insecticides on cottonalone and China which is one of the major consumers of insecticides on cotton are reportedly strongly considering the prospects of introduction of Bt transgenic cottons.

Cotton acreage in India is only 5% of the total cropped area, yet it consumes more than 50% of the total insecticides used in the country. The pattern of usage, however, is not uniformly, spread, for instance in Andhra Pradesh alone, where cotton cultivation is only 0.3% of the total cropped area of the country, accounts for 17% of the pesticide use on cotton in India. Again within the average picture of thestate, coastal AP uses 30% more pesticide than the state average. The approximate estimates of insecticide sales in Guntur district alone, where cotton is grown in 0.15 mHa, is about US $ M. Clearly areas with maximum pecticides use per hectare, necessitated mostly due to bollworms, are likely targeted market niches for the bollworm resistant transgenic cottons. Moreover, since the price of the transgneic seed in the US is nearly four times the high input cotton cultivators of the irrigated belt Growing Bt transgenics and herbicide tolerant transgenics on marginal land is not recommended by the company as these would not show any impact in the absence of strong insect pressure and broad leaved weeds. In most areas of rainfed cottons the use of insecticides and pest pressures are low hence, it may not be economical in such belts, to encourage the use of transgenics.

Now transgenic production technology for many crops has become an established routine procedure in many countries including India. The research on transgenic cotton in government funded research labs in India is almost in the final stages. Meanwhile Mahyco-Monsanto biotech is ready to market Bt transgenic cotton (bollgard) in India. The introduction of transgenics through these companies are to be in the form of FL seeds wherein Bt genes are inherited from the transgenic exotic germplasm. Such seeds may also be priced at a premium and it remains to be seen how enthusiastic the response of Indian farmers would be to the expensive input. Transgenic releases from government organizations would definitely enforce a competitive pricing to restrict the monopoly of the private companies.

The transgenic technology is also likely to be extremely beneficial to the private sector. For instance, a transgenic with resistance to a specific herbicide marginalises the use of other herbicides for chemical weed control and the chemical company holding the patent on the herbicide might be prove to be beneficial to the resource rich farmers. Clearly, here the targeted market is primarily high input farmers. The monopoly of transgenics due to patenting rights held by the biotech companies may also lead to high pricing of the seeds.

A Significant socioeconomic issue that arises from the introduction of transgenics into the Indian farming system is that the high priced seeds may benefits the prosperous and large farmers thus providing a negative externality on small and marginal farmers. On the other had it is also argued that the developments from the application of biotechnology would be beneficial to low input farming practises wherein the cost of chemical inputs can be minimised. It is now only a matter of time before we experience the full social, economic and environmental impact of transgenics in our country.

Benefits of Biotechnology in Agriculture

Everything in life has its benefits and risks, and genetic modification is no exception. Much has been said about potential risks of genetic modification technology, but so far there is little evidence from scientific studies that these risks are real. Transgenic organisms can offer a range of benefits above and beyond those that emerged from innovations in traditional agricultural biotechnology. Following are a few examples of benefits resulting from applying currently available genetic modification techniques to agricultural biotechnology.

Increased Crop Productivity

Biotechnology can help to increase crop productivity by introducing such qualities as disease resistance and increased drought tolerance to the crops. Researchers can select genes for disease resistance from other species and transfer them to important crops. For example, researchers from the University of Hawaii and Cornell University developed two varieties of papaya resistant to papaya ringspot virus by transferring one of the virus' genes to papaya to create resistance in the plants. Seeds of the two varieties, named 'SunUp' and 'Rainbow', have been freely distributed to papaya growers since May of 1998.

Further examples come from dry climates, where crops must use water as efficiently as possible. Genes from naturally drought-resistant plants can be used to increase drought tolerance in many crop varieties.

Enhanced Crop Protection

Farmers use crop-protection technologies because they provide cost-effective solutions to pest problems which, if left uncontrolled, would severely lower yields. For example, crops such as corn, cotton, and potato have been successfully transformed through genetic engineering to make a protein that kills certain insects when they feed on the plants. The protein is from the soil bacterium *Bacillus thuringiensis*, which has been used for decades as the active ingredient of "natural" insecticides.

In some cases, an effective transgenic crop-protection technology can control pests better and more cheaply than existing technologies. For example, with *Bt* bred into a corn crop, the entire crop is resistant to certain pests, not just the part of the plant to which *Bt* insecticide has been applied. In these cases, yields increase as the new technology provides more effective control. In other cases, a new technology is adopted because it is less expensive than a current technology with equivalent control.

There are cases in which new technology is not adopted because for one reason or another it is not competitive with the existing technology. For example, organic farmers apply *Bt* as an insecticide to control insect pests in their crops, yet they may consider transgenic *Bt* crops to be unacceptable.

Improvements in Food Processing

The first GMO food product to receive regulatory approval, in 1990, was *chymosin*, an enzyme produced by genetically engineered bacteria. It replaces calf rennet in cheese-making and is now used in 60 percent of all cheese manufactured. Its benefits include increased purity, a reliable supply, a 50 percent cost reduction, and high cheese-yield efficiency.

Improved Nutritional Value

Today there is increasing interest in improving the nutritional value, flavor, and texture of foods. Transgenic crops in development include soybeans with higher protein content, potatoes with more nutritionally available starch and an improved amino acid content, beans with more essential amino acids, and rice with the ability produce beta-carotene, a precursor of vitamin A, to help prevent blindness in people who have nutritionally inadequate diets.

Better Flavor

Flavor can be altered by enhancing the activity of plant enzymes that transform aroma precursors into flavoring compounds. Types of peppers and melons with improved flavor are currently in field trials.

Fresher Produce

Genetic modification can result in improved keeping properties to make transport of fresh produce easier, giving consumers access to nutritionally valuable whole foods and preventing decay, damage, and loss of nutrients. Transgenic tomatoes with delayed softening can be vine-ripened and still be shipped without bruising. Research is under way to make similar modifications to broccoli, celery, carrots, melons, and raspberry. The shelf-life of some processed foods such as peanuts has also been improved by using ingredients that have had their fatty acid profile modified.

Environmental Benefits

When genetic engineering results in reduced pesticide dependence, we have less pesticide residues on foods, we reduce pesticide leaching into groundwater, and we minimize farm worker exposure to hazardous products. With Bt cotton's resistance to three major pests, the transgenic variety now represents half of the U.S. cotton crop and has thereby reduced total world insecticide use by 15 percent! Also, according to the U.S. Food and Drug Administration (FDA), "increases in adoption of herbicide-tolerant soybeans were associated with small increases in yields and variable profits but significant decreases in herbicide use."

Benefits for Developing Countries

Genetic modification technologies can help to improve health conditions in less developed countries. Researchers from the Swiss Federal Institute of Technology's Institute for Plant Sciences inserted genes from a daffodil and a bacterium into rice plants to produce "golden rice," which has sufficient beta-carotene to meet total vitamin A requirements in developing countries with rice-based diets. This crop has potential to significantly improve vitamin uptake in poverty-stricken areas where vitamin supplements are costly and difficult to distribute and vitamin A deficiency leads to blindness in children.

Biotechnological Applications for The Improvement of Indian Tobacco

India is one of the top 10 countries cultivation tabacco for the part ten decades both for domestic and international market, India is the third largest producer of tobacco in the world with an output of approximately 520 m kg. Of which FCV tobacco account for 120 m kg. The economic importance of to Indian exchequer is peogressively increasing in the last 10 years. Tobacco generates 10% of total excise revenue and tobacco 4% of the value of total agricultural exports, India has the unique distinction of growing different types of tobacco in different Agroclimatic zones.

Tobacco provides direct and indirect lively-hood to 26 million in India including 6 million farmers and workers. Many small and marginal farmers were benefitted by growing tobacco. Research efforts to identify alternative crops to tobacco indicated that no other crops is as remunerative as tobacco. Stability of the worldwide demand of the crop, the crops hardiness and ability to grow in clumate and soils non-suitable for other crops, relative ease in transporting tobacco and tobacco's profitability compared to other crops are incentives for farmers growing tobacco thus making tobacco one of the most appealing cash crops to farmers.

Tobacco crops is considered as major employer worldwide. Despite technological progress tobacco cultivation remained as one of the most labour intensive crops among all other agricultural crops. Women in particular and skilled labourers, who would otherwise have little chance of employment are able to earn living by working in tobacco cultivation. WHO in its report has acknowledged that in developing countries tobacco can create vast employment. In view of its great role in the economy of the country and farmers, further improvement of tobacco crop through Biotechnological tools will benefits them a great deal.

Biotechnological Applications

Scientific system of tobacco cultivation in India was introduced with the establishment of Central Tobacco Research Institute (CTRI) at Rajahmundry in 1947. Tobacco research in India over the past 50 years through CTRI has made significant contributions to the economic upliftment of tobacco farmers by identifying high yielding varieties and suitable package of practices This is reflected in the good reputation for Indian Tobacco in the International market, as leaf of desired quality traits as per the specific requirement of traders is produced.

Initially, conventional breeding methods were utilised for improving the yield and quality of tobacco. However, later application of biotechnology tools for supplementing the conventional tobacco improvement programmes for initiated. Different biotech techniques like anther culture, fertilized ovule culture micropropagation, protoplast culture somatic hybridization and also genetic engineering are being utilised. These techniques are helpful is strengthening the conventional breeding programmes where ever they are week and failing to serve the purpose. Progress of biotechnology world, in different aspects, at CTRI is given follow:

Yield Improvement

Yield improvement through conventional breeding programmes takes fairly long time. This can be reduced substantially (3-4 years) by culturing anthers (Anther culture). Of elite lines from early breeding populations in tissue culture. Through anther culture several dihaploids are produced and evaluated every year. An improved line D1 was developed, in this way, which is superior in yield. Efforts to incorporate budworm resistance through semicolon variation in this line is in progress. Stable resistant D1 someclones are there in field testing stage.

Resistance to Biotic and Abiotic Stresses:

Wild species of tobacco possesses resistance to various tobacco pests and disease. It is difficult and in some cases impossible to transfer this resistance from wild species to cultivated tobacco through conventional methods. In such cases, interspecific hybrids were rescued through fertilized ovule culture and mass multiplied for further testing by micropropagation, In this way, interspecific hybrids possessing resistance to leaf spot diseases, black shank, root knot nematodes leaf eating caterpillar, aphids, budworm etc. were developed and evaluated for their resistance. Thus, stable pest and disease resistant lines were identified through Biotechnological tools. Efforts to transfer desirable characters from wild species through somatic hybridization is in progress.

Somaclonal variation was attempted to fortuity tobacco lines resistance to white fly (Bemisia tabci Genn.) (leaf curl) and budworm (Heliothis armigera Hb) and the results are encouraging. Somaclonal variation is highly effective to create variation when desired variation is impossible to get through conventional breeding methods and also variation is required only for one character.

Transgenic Bt tobacco cultivars processing resistance to leaf eating caterpillar (Spodoptera litura F) and budworm (Heliothis armigera Hb) were developed. Production of transgenics will give scope for transferring desirable characters from different species and genera.

Preliminary attempts were made to screen lines for drought tolerance in tissue culture. Such efforts will save expenditure on extensive field evaluation and work can be carried out in the offseason also. Transgenic tobacco having wheat desiccation tolerant protein were evaluated water stress tolerance.

Herbicide Resistance

Orabache cernua os a root parasite of tobacco causing heavy loss in yield and quality. Glyphosate at lower does was found to kill Orabanche, but it is phytotoxic to tobacco. Hence research on production of herbicide (glyphosate) resistant tobacco lines through tissue culture was initiated. When farmer grows herbicide resistant lines in the field he can use berbicides without any harmful effect on tobacco, for the effective control of weeds and Orabanche.

Conservation of Germplasm

Germplasm is the valuable source for various characters that are useful in future breeding programmes. This can also be conserved under in vitro conditions for long time. Also lines that cannot be maintained under field conditions due to non-flowering and sterility can be effectively maintained under tissue culture conditions. Interspecific F1s, mamoth mutants, in vitro rescued hybrids protoclones, BT transgenics various mutants are routinely being maintained.

Future Thrusts

Biotechnology can help Indian tobacco farmer in many ways, Particularly in tobacco improvement research and seed purity disputes, if offers a large scope. Research on tobacco improvement benefit farmer as he can get high yielding and pest and disease resistant tobacco for cultivation. In research various aspects like in vitro conservation and molecular characterisation of germplans, isolation, characterisation and transfer of desirable genes from across genera, molecular aided selection in breeding programmes, semicolon variation can play a great role. Utilising Biotechnological techniques tobacco lines that can yield higher amounts of phytochemicals which are useful in the production various drugs and pesticides can also be developed.

In recent times many disputes about seed purity among farmers are coming up. At present morphological criteria is only being used for varietal identification. As morphology is highly subjective in nature and also varies with the type of environment the plant is grown, it is becoming highly difficult to solve the disputes. Under this situation molecular characterisation of cultivated varieties will help to identify correct varieties beyond doubt.

In future, giving emphasis on the above said lines will go a long way in benefitting the India tobacco crop as well as the farmer. However, as biotechnology got wider applications, any thing that put the farmer and environment at disadvantage need to be discourages.

Agricultural Biotechnology: Adoption of Biotechnology and its Production Impacts

Driven by farmers' expectations of lower production costs, higher yields, and reduced pesticide use, the rate at which U.S. farmers adopt genetically engineered (GE) crop varieties has jumped dramatically. It has been estimated that about 220 million acres of GE crops with herbicide tolerance and/or insect resistance traits were cultivated worldwide in 2005, an 11-percent increase over acreage in 2004, and U.S. acreage accounts for about 55 percent of this. However, actual benefits in terms of costs, yields, and pesticide use vary with the crop and engineered trait.

Adoption of herbicide-tolerant (HT) crops, which carry genes that allow them to survive certain herbicides that previously would have destroyed the crop along with the targeted weeds, has been particularly rapid since they first became available to farmers in 1996. HT soybeans expanded to 89 percent of U.S. soybean planted acreage and HT cotton expanded to 65 percent of cotton acreage in 2006. Adoption of insect-resistant (Bt) crops, containing the gene from a soil bacterium Bacillus thuringiensis (Bt), has also expanded. Use of Bt cotton reached 57 percent of planted cotton acreage in 2006 and Bt corn use grew from about 1 percent of corn acreage in 1996 to 40 percent in 2006.

Factors Affecting Farmers' Adoption of GE Crops

According to surveys conducted by USDA in 2001-03, most farmers (59-79 percent) adopting GE corn, cotton, and soybeans indicated that they did so mainly to "increase yields through improved pest control." The second most cited aim was to "save management time and make other practices easier" (15 to 26 percent, except for Bt corn, which was much lower); the third reason was to "to decrease pesticide costs" (9-17 percent of adopters). All other reasons combined accounted for 3-7 percent of adopters. Hence, factors expected to increase economic profitability by increasing revenues per acre (yield times price of the crop) or reducing costs (operator labor, pesticides) are expected to promote adoption most.

Effect of GE Crops on Yields

It is difficult to estimate the farm-level effect of genetically engineered crops on yields because impacts vary with the crop and technology examined. Yields also depend on locational factors such as soil fertility, rainfall, and temperature, which can also influence the very presence of pests.

An additional problem with estimating the benefit of GE crops is self-selection: farmers are not assigned randomly to the two groups (adopters and nonadopters) but make the adoption choice themselves. Therefore, adopters and nonadopters may be systematically different and these differences may manifest themselves in farm performance and could be confounded with differences due purely to adoption. GE crops do not increase the yield potential of a hybrid. In fact, yield may even decrease if the varieties used to carry the herbicide-tolerant or insect-resistant genes are not the highest yielding cultivars. However, by protecting the plant from certain pests, GE crops can prevent yield losses compared with non-GE hybrids, particularly when infestation of susceptible pests occurs.

This effect is particularly important in the case of Bt crops. Before the commercial introduction of Bt corn in 1996, the European corn borer was only partially controlled using chemical insecticides. The economics of chemical use were not always favorable and timely application was difficult. For these reasons, farmers often accepted yield losses (of 3-6 percent per one corn borer per plant, depending on the stage of plant development) rather than incur the expense of chemical pesticides to treat the insect. An ERS study estimated the impact of adopting GE crops using 1997 survey data. Herbicide-tolerant soybeans and cotton and Bt-enhanced cotton were modeled individually. In each model, pest infestation levels, other pest management practices, crop rotations, tillage, and self-selection were controlled for statistically. Geographic location was included as a proxy for soil, climate, and agricultural practice differences that might influence impacts of adoption.

Results of such modeling can be interpreted as an elasticity or responsiveness to the change in a particular impact (yields, pesticide use, or profits) relative to a small change in adoption of the technology from current levels. The results can be viewed in terms of aggregate impacts across the entire agricultural sector as more producers adopt the technology, or in terms of a typical farm as they use the technology on more of their land. As with most cases in economics, the elasticities estimated in the quantitative model should only be used to examine small changes (say, less than 10 percent) away from current levels of adoption.

The study shows that adoption of herbicide-tolerant cotton led to significant increases in yields. The elasticity of yields with respect to the probability of adoption of herbicide-tolerant cotton is +0.17. That is, an increase of 10 percent in the adoption of herbicide-tolerant cotton led to a 1.7-percent increase in yields. Similarly, the adoption of Bt cotton in the Southeast increased yields significantly (elasticity of yields is +0.21). On the other hand, increases in adoption of herbicide-tolerant soybeans led to small (but significant) increases in yields (elasticity of yields is 0.03).

Effect of GE Crops on Pesticide Use

On the environmental side, pesticide use on corn and soybeans has declined since the introduction of GE corn and soybeans in 1996.

In addition, ERS research suggests that, controlling for other factors, pesticide use declined with adoption. The overall reduction in pesticide use associated with the increased adoption of GE crops (Bt cotton; and HT corn, cotton, and soybeans, using 1997/1998 data) also resulted in a significant reduction in potential exposure to pesticides. The decline in pesticide applications was estimated to be 19.1 million acre-treatments (Fernandez-Cornejo and McBride, 2002). Total pesticides applied to corn, soybeans, and cotton declined by about 2.5 million pounds (active ingredients), despite the (slight) net increase in the amount of herbicides applied to soybeans.

Crop Breeding, Genetics & Cytology

Phenotypic stability has long been recognized as an important target in plant breeding. Stability is influenced in part by the genetic structure, i.e., level of heterogeneity and heterozygosity, of the cultivar. Yet, very little is known about the genetic components underlying stability, and how population improvement strategies influence stability. We examined 12 maize (*Zea mays* L.) breeding populations selected via reciprocal recurrent selection (RRS), selfed progeny recurrent selection (S), or a method combining RRS and S (COM), to examine changes in the genetic structure of the phenotypic stability of three traits (grain yield, grain moisture, and broken stalks), and two associated selection indices. Partitioning of the genotype x environment sums of squares from diallel matings of the original (C_0) and advanced (C_A) cycle populations into linear trends indicated that only grain yield and the unadjusted performance index (UPI) followed a predictable linear response. Grain yield and UPI linear trends were further partitioned by Gardner and Eberhart Analysis III to examine the genetic components of stability. We found that recurrent selection (RS) improved grain yield stability, and that this trait is heritable, predictable, and mostly controlled through additive gene action. Improvement in grain yield stability was observed both in cross and per se performance and was accompanied by significant improvement in the mean performance of the populations. However, the improvement in grain yield stability did not result in substantial changes in the general combining ability (g_i) estimates of most populations. Our results indicate that grain yield stability can be improved through RS by selecting solely for mean performance across multiple environments.

Quantitative genetic theory, states that an individual's phenotype (P) is the product of the genetics (G) of the individual, the environment (E) that the individual is exposed to, and the interaction that occurs between the genotype of the individual and the environment (G x E). Large G x E effects tend to be viewed as problematic in breeding because the lack of a predictable response hinders progress from selection (Dudley and Moll, 1969). This idealized predictable response across multiple environments is generally referred to as stability (Cannon, 1932). When stability concepts are applied to breeding–selection for increased grain yield, the idealized genotype is one that is capable of utilizing the resources available in higher yielding environments and has a mean performance that is above average in all environments.

Yield stability is influenced in part by the genetic structure of the variety. More heterozygous varieties and more heterogeneous varieties are less affected by environmental differences (Lerner, 1954; Lewontin, 1957; Allard and Bradshaw, 1964). For example, maize double-cross hybrids have smaller G x E interactions and are more stable than maize single-cross hybrids (Sprague and Federer, 1951). More homogeneous generations, inbred and F_1, have larger G x E interactions than more heterogeneous generations, F_2 and BC_1, and a more homozygous generation, inbred, has a larger G x E interaction than a more heterozygous generation, F_1 (Valdivia-Bernal and Hallauer, 1991). To reduce the complications that the G x E interaction creates when selecting superior genotypes, many attempts have been made to (i) understand the environmental components causing the G x E interaction (Epinat-Le Signor et al., 2001; Jeutong et al., 2000), (ii) examine the G x E interaction biometrically (Lin et al., 1986; Zobel et al. 1988; Finally and Wilkinson, 1963; Yan et al., 2001), and (iii) develop selection strategies that involve a stability parameter (Magari and Kang, 1993). However, very little effort has been focused on examining the genetic components underlying, in particular, yield stability. In maize, the limited information regarding the genetics of yield stability comes from diallel studies involving inbred lines. These results indicate that stability (both positive and negative) is heritable and controlled by additive gene action (Eberhart and Russell, 1966, 1969).

One approach to examining stability is to further partition the G x E interaction from a traditional analysis of variance (ANOVA) into linear trends and a departure from linear (residual) (Finally and Wilkinson, 1963; Eberhart and Russell, 1966). Linear trends can then be further partitioned by Analysis III of Gardner and Eberhart (1966) to arrive at estimates of genetic effects. It is then possible to examine the genetic structure underlying yield stability in terms of heterosis, general combining ability (GCA), and specific combining ability (SCA). The linear portion of the G x E interaction presumably has an additive and a nonadditive genetic component. Understanding the relative importance and genetic nature of these components and how selection influences them will greatly enhance our knowledge of yield stability and will help determine if selection strategies designed to enhance stability are important in breeding programs. In this paper, we present the results from two diallels involving, respectively, the originating gene pool (C_0) and the most advanced cycle (C_A) of 12 breeding populations. Specifically, we were interested in examining (i) if genetic progress for yield stability has been achieved indirectly through RS, (ii) if RS produced changes in the genetic structure of the populations for yield stability, and (iii) the underlying genetic components that influence yield stability.

Material and Methods

Genetic Materials

The 11 C_0 and 12 C_A (C_2–C_6) early (2550–2900 Ontario Crop Heat Units [OCHU] [Brown and Bootsma, 1993]) University of Guelph breeding populations in this study included CG-SynA (S), CG-SynB (S), CG-Wigor (RRS) and CG-SynA (RRS), CG-HOPE Elite A (RRS) and CG-HOPE Elite B (RRS), CG-Stiff Stalk (COM) and CG-Lancaster (COM), CG-CBI (RRS) and CG-CBII (RRS), and CG-Cross Canada Gene Pool A (RRS) and CG-CCGP B (RRS). The two CCGP populations were selected from a common C_0, resulting in 11 C_0 populations and 12 C_A populations. All other populations were unrelated except that CG-SynA (RRS) and CG-SynA (S) originated from different cycles of selection in the CG SynA gene pool. Four to six cycles of RS had been completed in each population except for the CCGP populations, which had undergone two cycles of RS. For each cycle, families within a population were selected visually during development and subsequently in replicated trials in two environments, i.e., locations. Further information regarding the populations can be found elsewhere (Lee, 2002). Unadjusted (UPI) and adjusted (API) performance indices were used as the selection criteria for advancement until the populations had attained a level of maturity comparable to adapted germplasm for 2600 OCHU. The UPI was the ratio of grain yield to the percentage of grain moisture at harvest, while the API was the same ratio, but based on standing plants only, i.e., whole-plot grain yield x the percentage of plants neither lodged nor with broken stalks. Progenies were ranked by each index and selected based on the sum of ranks to identify those progenies for intercrossing to form the next cycle. When satisfactory grain moisture at harvest was reached, selection was based only on the sum of ranks for grain yield and grain yield of standing plants. In general, 20 selected families were intercrossed to form a new cycle.

In 1999, separate half-diallel matings, not including reciprocals, were made with the C_0 populations ($n[n - 1]/2 = 55$ crosses) and C_A ($n[n - 1]/2 = 66$ crosses). For each cross, six ears were pollinated with bulked pollen from six plants from the other population. Reciprocal crosses were then bulked, resulting in a total of 12 ears per cross. Individual plants were used as either males or females in the crosses, but not both, for a total of 24 S_0 plants, representing 24 gametes from each population. This approach samples predominant alleles and linkage blocks; rare alleles and linkage blocks are likely not represented in the population crosses and is similar to other studies utilizing population crosses (Moll et al. 1977; de Leon and Lonnquist 1978; Douglas et al. 1961; Gouesnard et al. 1991). Seed was increased for each of the 11 C_0 and 12 C_A parental populations via chain sobbing and at least 70 individual ears were harvested from each population. Equal amounts of seed from each ear of the populations per se were bulked for use in the experimental trials.

Experimental Procedure and Data Collection

The experiment included 144 entries, 66 crosses from the C_A diallel mating, 55 crosses from the C_0 diallel mating, 12 advanced populations per se, and 11 C_0 populations per se. Yield trials were grown in both 2000 and 2001 at three southwestern Ontario locations (Alma [2500 OCHUs], Elora [2600 OCHUs], and Woodstock [2850 OCHUs]) in a 12 x 12 partially balanced lattice design with three replications at each location. The soil type at all locations is Guelph loam (Typic Hapludalf). Fertilizer (N, P_2O_5, and K_2O, respectively) was applied on the basis of soil tests at the rate of Alma (2000) 155, 58, and 74 kg ha^{-1}; Alma (2001) 122, 75, and 57 kg ha^{-1} supplemented with 33 690 L ha^{-1} liquid swine manure; Elora (2000) 140, 50, and 50 kg ha^{-1} and 135, 32, and 76 kg ha^{-1}; Elora (2001); Woodstock (2000) 150, 66, and 108 kg ha^{-1}; and Woodstock (2001) 167, 59, and 75 kg ha^{-1}. Weeds were controlled by conventional herbicides. Experimental units were two-row plots, 5.78 m long, with a spacing of 0.76 m between rows. Trials were overplanted and thinned at the six-leaf stage to uniform stands of 68 000 plants ha^{-1} (60 plants $plot^{-1}$). All trials were machine planted (New Idea four-row planter New Idea Company, Cold Water, OH) in May with Almaco planting cones and machine harvested (Almaco SPC40-2 two row combine [ALMACO, Allan Machine Company, Nevada, IA)] in October or November. Three traits and two indices were measured, respectively: machine-harvested grain yield (kg ha^{-1}) adjusted to 155 g kg^{-1} grain moisture grain moisture at harvest, percentage of broken stalks (plants broken below the ear or inclined more than 45° from the vertical), UPI, and API. All five parameters were analyzed and will be referred to as traits.

Results and Discussion

Analysis III

Environments, genotypes, and G x E were significant sources of variation for all traits (Doerksen et al., 2003). Genotype x environment was further partitioned into linear trends and deviations from linear. The linear trend was not significant for grain moisture, broken stalks, or API. Grain yield and UPI were the only traits where both linear trends and deviations from linear trends across environments were significant. However, as a percentage of the G x E sums of squares, linear trends for grain yield and UPI accounted for only 25 and 26% of the variance, respectively. Linear trends were further partitioned into C_A vs. C_0, E x C_0–linear, and E x C_A–linear. The C_A vs. C_0 contrast was significant for both grain yield and UPI, indicating significant differences in phenotypic stability between C_0 and C_A. Furthermore, there were significant differences among regression coefficients within the E x C_0–linear and within the E x C_A–linear, indicating that differences in phenotypic stability exist between populations.

The E x C_A–linear and E x C_0–linear variances for grain yield and UPI were further partitioned into parents, heterosis, and crosses by Analysis III (Gardner and Eberhart,

1966). In both C_A and C_0, differences in linear trends were largely due to crosses (88 and 83%, respectively). Surprisingly, no significant heterotic effect among regression coefficients was detected, implying that heterosis in this set of material is not conferring increased phenotypic stability. This is contrary to other studies that have suggested that heterosis is important for conferring stability (Valdivia-Bernal and Hallauer, 1991). Heterosis is important for grain yield in this set of materials. When the entry variance was partitioned in a similar manner, a significant heterosis effect was observed for grain yield in both C_0 and C_A (Doerksen et al., 2003). But, perhaps there is sufficient heterogeneity and heterozygosity in C_0, which is maintained in C_A, that additional heterozygosity does not confer additional stability. Crosses were further partitioned into GCA and SCA for both E x C_A–linear and E x C_0–linear. Differences in linear trends due to GCA were significant in both C_0 and C_A, while differences due to SCA were significant only in C_A. *F* tests of the ratio of GCA to SCA mean squares were significant in C_0 and C_A for grain yield (3.03 and 2.57, respectively), but not for UPI (1.99 and 1.87, respectively), indicating that grain yield stability is mostly controlled by additive genetic effects while UPI stability is controlled both by additive and nonadditive effects. Yield stability is heritable and predictable as indicated by significant GCA effects in both the C_0 and C_A. Yield stability has been improved through recurrent selection indicated by the significant C_A vs. C_0 contrast. In both gene pools, i.e., C_0 and C_A, there appears to be sufficient heterozygosity and heterogeneity that heterosis does not confer any additional stability. And finally, mostly additive genetic affects are controlling grain yield stability.

Stability Analysis

Yield stability was examined initially by AMMI analysis on the C_0 and C_A separately and combined. The first two principal components accounted for 41 and 20% of the variation, respectively. However, large environmental scores and small genotype scores were obtained, resulting in a tight clustering of the genotypes at the center of the plot (data not shown). Because of the lack of informativeness of the AMMI results, the LR approaches of Finally and Wilkinson (1963) and Eberhart and Russell (1966) were used to examine the mean response of a population both in crosses and per se. Linear regression, a Type 2 stability statistic (Lin et al., 1986), was used because we were only interested in comparisons among the set of genotypes, and the LR approach allowed us to exploit the genetic structure that exists among the entries.

Because of the additive nature (significant GCA and lack of heterosis) of grain yield stability observed with Analysis III, we used, as a measure of a population's stability, the mean b_i values for a population averaged over all crosses involving that population. As implied by Analysis III, all C_A populations in crosses had significantly higher b_i values than their corresponding C_0 populations, showing that RS had improved the responsiveness of the populations to higher yielding environments. Accompanying this increase in responsiveness was also a significant increase in mean yield in crosses for all populations. Taken together, it appears that the responsiveness did not come at the expense of average performance.

Relationship between C_0 and C_A Genetic Effects for Yield Stability

In C_0, the differences among linear trends for g_i estimates were largely due to a large negative contribution from Wigor (-0.31) and a large positive contribution from SynA(S) (0.19). This indicated that Wigor and SynA(S) were contributing lower and higher grain yield to their crosses, respectively, in higher yielding environments. In C_A, the differences among linear trends for g_i estimates were largely due to large negative contributions from Wigor (-0.24) and SynA(S) (-0.19). The significant rank correlation (r = 0.59, P > 0.05) between C_0 and C_A g_i estimates suggested that the stability parameter, while improved on through recurrent selection, was essentially a characteristic present in the original population.

Consequences of Recurrent Selection in the 12 Breeding Populations

We have now examined two important aspects of the 12 RS populations. First, we were interested in what changes in genetic structure had occurred that influenced grain yield (Doerksen et al., 2003). We found that genetic improvement occurred in both the per se and cross performance of most populations. Accompanying the favorable changes in population performance were less favorable shifts from predominantly additive genetic effects in C_0 to greater nonadditive genetic effects in C_A. This shift did not substantially change g_i estimates of most populations. However, in the case of grain yield, the underlying components of g_i effects were altered in their relative importance. General combining ability effects in C_0 were caused primarily by population per se effects (v_i), while in C_A, the GCA effects were caused predominately by parental heterotic effects (h_i) (Doerksen et al., 2003).

The other aspect of these populations, addressed in this paper, is phenotypic stability for grain yield. Yield stability is heritable and predictable. It has also been improved through recurrent selection. In both gene pools, i.e., C_0 and C_A, there appears to be sufficient heterozygosity and heterogeneity that heterosis (i.e., population crosses) does not confer any additional stability. Mostly additive genetic affects are controlling grain yield stability, with no apparent shift toward more nonadditive effects in C_A. Recurrent selection for only grain yield across multiple environments has resulted in improved cross yield stability while simultaneously improving mean performance of the populations. It is important to note that the number of test environments during RS was only two per cycle, i.e., four cycles = eight performance tests, six cycles = 12 performance tests. Apparently, the two environments per cycle, plus the different years of testing each cycle, even though a relatively small number (eight to 12 for most populations), has been sufficient to select (indirectly) for stability of grain yield. Or, interpreted another way, the reduction of deleterious gene frequencies for grain yield, which would be especially effective in the early cycles of RS, develops a more refined heterogeneous, heterozygous population, and therefore a more stable one. In effect, a good *yield gene* is also a good *stability gene*.

We also can draw some preliminary conclusions regarding the effectiveness of the various methods of improvement, S, RRS, and COM. One of the most surprising observations from the diallels is that in many cases in C_A, the crosses involving the RRS pairs were not superior to crosses involving other populations. And in some instances, Wigor-SynA(R) and Stiff Stalk-Lancaster, most of the crosses involving other populations were superior to the RRS pair cross. This is contrary to RS theory. Given this set of data, it appears that RRS will not result in population pairs that are consistently superior to populations improved independently from one another.

Biotechnologycal Application: Releavance to the Indian Farmer in the Field of Voat and Sheep Rearing

In this field biotechnological applications are extremely relevant if we are to made rapid progress in raising the productivity of these small ruminants.

Techniques such as embryo transfer, dilution of semen and its cryo preservation and artificial insemination (A.I.) are well developed and expertise in these fields is available. This technology is extensively used in the case of cows and to some extent Buffaloes.

The main lacuna in the case of sheep and goats is the non availability of improved breeds which could be utilised. As these animals are owned by the poor strata of society and there is a mistaken prejudice that they are responsible for environmental degradation no sustained efforts have been made to improve indigenous breeds and thus make available genetic stock for upgrading. In spite of this the numbers of these animals have been increasing rapidly (*2.5% per year in the case of goats) due to an increasing demand for meat. Our breeds have developed over the centuries to survive and adapts themselves to harsh environmental conditions but do not universally possess ability to respond to better management and nutrition.

Outstanding work in the case of the Sirohi goat was carried out from 1981 to 1993 by the Indo-Swiss Goat Development and Fodder Production Project in Rajasthan where they showed that outstanding animals are available within this breed. However, before the impact of this project could be felt the project was discontinued.

Goats

Out institutes is a very small NGO with only three technically qualified staff, one of who is in an honorary capacity and one on deputation from the Government of Maharashtra. We decided to initially concentrate on crossbreeding using the outstanding South African Boer Goat the worlds only meat goat as an improve breed. With assistance from the Government of Australia we imported 20 embryos of this breed from Australia implanted them in local goats and now have a small flock of Boer goats.

This goat unlike many temperate breeds of Dairy goats have proved to be extremely hardy, producing excellent crossbreed when crossed with local goats. In order to utilise this germplasm effectively we have developed and modified the techniques of semen cryopreservation so that it can be done with low cost equipment producing pellets instead of straws. We have also improvised. A.I,. techniques using cervical insemination to get better conception rates. We have so far trained over 50 trainees among them farmers. We have supplied pure Boer goat frozen semen to Governmental agencies in Orissa, Tamil Nadu as well as to may private goat rearers in Maharashtra.

Our veterinarian is capable of carrying out multiple ovulation and embyo transfer if there is need for rapid expansion of our flock. However, this technology is expensive and only to be used if costs justify its use.

A I technology in goats has been spread to near by villages where fresh diluted semen as well as frozen semen have been used to produce 800 crossbreeds over the past five years. This technology is viable and easily transferable but unfortunately dependent on the sincerity of the inseminator.

Sheep

We were unable to identify a suitable improver breed to be imported which had the most desirable but greatly neglected attribute of prolificacy. We did locate a prolific and unusual breed in the hot humid sundarbands of West Bangal called the Garole. Up to 1993 this breed had remained unnoticed by the community of sheep breeders. We did however import a small flock of Awassi Dairy Sheep from Israel. This breed has produced an outstanding crossbred animal when crossed with the Malpura of Rajasthan. This crossbred has been developed by the Central Sheep and Wool Reserach Institute (CSWRI) using Awassi rams obtained from us.

The Australian Center for International Agricultural Research has sponsored a collaborative project between the University of New England, Australia, The national Chemical Laboratory, Pune and the Nimbkar Agricultural Research Institute (A SISTER NGO of Maharashtra Goat and sheep Research & Development Institute), Phaltan entitled," Prolific Worm Resistant Meat Sheep for Maharashtra, India.

This project is making full use of Biotechnology. We are comparing three breeds of Indian sheep the Grole, Deccani and the bannur. We hope to evaluate lamb production including prolificacy and resistance to parasites and try and determine the genetic basis of these differences by DNA analysis. We also hope to account for the interaction between measurement of resistance to parasistic nematodes and environmental effects. The project has completed one year and the results so far are encouraging.

Our ultimate objective is to provide shepherds with rams incorporating the prolificacy of the Garole, the growth potential of the Awassi and the meat quality and hardiness of the Bannur.

A very important but at present neglected application of Biotechnology which requires considerable investment is in the field of disease diagnosis and vaccine preparation. Once productive animals that respond to inputs are available and the rearers can afford to pay for the above services I am confident that the private sector will enter this field to provide this input.

It is imperative to review the cost benefit ratios of Government institutions and their contribution to the Goat and Sheep industries I think you will find that funds given to NGO's are more productive. There is much talk of helping NGO's but real assistance only seems to come from foreign aid agencies which is unfortunate.

Interspecific Hybridization as a Potential Method for Improvement of *Agrostis* Species

Interspecific and Intergneric hybridization between crop species and wild relatives followed by backcrossing is commonly used to introgress desirable genes into the crop (Stalker, 1980). This method has been used in breeding improved cultivars of numerous species (Kalloo, 1992). Such wide hybridization has also been used to generate new crop species, the best known being triticale (Lukaszewski and Gustafson, 1987). Interspecific hybridization has not, however, been widely utilized by turfgrass breeders (Brilman, 2001) and may offer new opportunities for cultivar improvement. In fact, many turfgrass species have evolved through natural interspecific hybridization (Cassler and Duncan, 2003).

Some *Agrostis* spp. are important turfgrasses and are used extensively on golf courses. Creeping bentgrass is the major species used on golf greens and fairways in temperate climates (Warnke, 2003). Colonial bentgrass and velvet bentgrass are also occasionally used (Brilman, 2003; Ruemmele, 2003). Each species has traits such as susceptibility to particular diseases and to drought and heat stress, which are being addressed in current academic and commercial breeding programs (Meyer and Belanger, 1997). Improvements in some of these traits may be possible through interspecific hybridization.

Hybrids between various *Agrostis* spp., identified as having intermediate characteristics, have been reported to occur in nature (Stuckey and Banfield, 1946; Bradshaw, 1958; Edgar and Forde, 1991). The most extensive investigation into interspecific hybridization among *Agrostis* spp. was done in the 1950s by Jones. Hybrids among creeping bentgrass, colonial bentgrass, and *A. gigantea* Roth (redtop bentgrass) were examined for chromosome pairing at metaphase 1 of meiosis (Jones, 1956a,b,c). Jones proposed a model for genome organization in these *Agrostis* spp.

Some *Agrostis* spp. were considered to be allopolyploids having some ancestral genomes in common.

The previous work established that interspecific hybrids between *Agrostis* spp. can occur, but there was no information on the frequency of hybridization. Identification of interspecific hybrids relied on subjective assessment of intermediate morphological characteristics and cytological analysis. Current molecular techniques utilizing transgenic plants and molecular markers provide a simpler and more precise method of hybrid identification.

The objective of this study was to investigate the frequency of interspecific hybridization between creeping bentgrass and colonial bentgrass, velvet bentgrass, redtop bentgrass, and dryland bentgrass (*A. castellana* Boiss. and Reut.) under optimum conditions. Dryland bentgrass and redtop bentgrass are used as low maintenance turfgrasses and seed of both is commercially produced (Brede and Sellmann, 2003). The chromosome numbers of the *Agrostis* spp. investigated range from 14 to 42. To obtain an assessment of the maximum biological potential for hybridization, flowering times were artificially manipulated and crosses were done under greenhouse conditions. Transgenic creeping bentgrass plants expressing the *bar* gene (Hartman et al., 1994; Lee et al., 1996), which confers resistance to the herbicide glufosinate [2-amino-4-(hydroxymethyl phosphinyl) butanoic acid], were used as the pollen parents. Transfer of the herbicide resistance transgene to progeny harvested from the nontransgenic *Agrostis* spp. was used to identify hybrids. Here we report on the frequency of hybridization and the fertility of the hybrids recovered.

Material and Methods

Plant Materials

Two lines of transgenic creeping bentgrass plants previously generated by particle bombardment (4475) and protoplast transformation (5061) were used as pollen parents (Hartman et al., 1994; Lee et al., 1996). These transgenic lines were previously found to produce viable pollen and their progeny segregated 1:1 for the *bar* gene (Lee et al., 1997; Belanger et al., 2003). The plants were clonally propagated in the greenhouse.

Velvet bentgrass seed was from the Western Regional Plant Introduction Station, Pullman, WA (PI 189141). Dryland bentgrass cv. Highland and redtop bentgrass cv. Streaker were from Great Western Seed Company, Albany, OR. Colonial bentgrass was from the Western Regional Plant Introduction Center (PI 171470, PI 172698, PI 252045, PI 494120) and from Seed Research, Inc., Corvallis, OR.

Individual seedlings of the nontransgenic *Agrostis* spp. were established in the greenhouse. *Agrostis* spp. require vernalization for flower induction. In July 1997, a nursery plot of the transgenic creeping bentgrass and the nontransgenic *Agrostis* spp. was established for use in crossing in spring 1998.

Crosses and Screening for Hybrids

There is variation in the seasonal timing of flowering in different species of *Agrostis*. To ensure that flowers of comparable developmental stages would be available from the various species, flowering was induced in some plants of each species by moving them into a containment greenhouse maintained under long days (18 h). Beginning in April 1998, individual plants of each species were transferred to the containment greenhouse at weekly intervals. The temperature during the day was 18 to 21°C and at night was 15 to 18°C.

For crossing, panicles of unopened flowers of a transgenic and nontransgenic plant of similar developmental stage were enclosed in bags constructed from interfacing material obtained from a local fabric store. Only the panicles from the nontransgenic parent were harvested. For each cross, the nontransgenic parent plant originated from a single seed and thus represents a unique genotype. Ten to 13 different nontransgenic plants for each species were used in the crosses. By setting up multiple crosses we could be assured of testing the hybridization potential of different genotypes within each species. Seeds obtained from the crosses were surface sterilized and plated on solidified Murashige and Skoog basal medium (Gibco BRL, Gaithersburg, MD) supplemented with 1 mL L^{-1} 1000X Gamborg's Vitamin Powder (Sigma, St. Louis, MO), 100 mg L^{-1} myo-inositol (1,2,3,4,5,6-cyclohexanehexol), and 30 g L^{-1} sucrose (1-alpha-D-glucopyranosyl-2-beta-D-fructofranoside) in 100- by 25-mm Petri dishes. The dishes were placed at 4°C for 1 wk and then moved to a lighted growth chamber at 25°C for germination. When the plants were 2- to 3-cm tall they were transplanted into 96 well flats in the greenhouse. After 4 wk of establishment in the greenhouse, the plants were sprayed with the commercial herbicide Finale (AgrEvo Environmental Health, Montvale, NJ) at a final concentration of glufosinate of 0.35%. Plants that survived the spray were reported and maintained in the greenhouse. The surviving plants were later sprayed again to confirm that they were herbicide resistant.

Polymerase Chain Reaction Analysis

Polymerase chain reaction (PCR) analysis was used to confirm the presence of the *bar* gene in some of the interspecific hybrids. DNA was isolated from leaf blades as previously described (Reddy et al., 1996). The primers used were designed to amplify a 446-base-pair region of the *bar* coding sequence. The primer sequences were 5'TGCACCATCGTCAACCACTACATC3' and 5'AGGCTGAAGTCCAGCTGCCAGAAA3'. Polymerase chain reactions were performed in a total volume of 50 μL. Each reaction contained 1 U of Elongase Enzyme Mix (Life Technologies, Gaithersburg, MD), 20 pmol of each primer, 200 μM deoxynucleotide triphosphates, 1.4 mM $MgCl_2$, and 100 ng of total genomic DNA. Polymerase chain reaction cycling parameters were 35 cycles of 30-s denaturation at 94°C, 45-s annealing at 60°C, and 120-s extension at 68°C. Polymerase chain reaction products were analyzed by electrophoresis in 1% (w/v) agarose gels.

Hybrid Fertility

To assess the fertility of the recovered hybrids, a nursery plot of the hybrids was established in September 1999 to allow vernalization through the winter. In the spring of 2000, three individuals from each hybrid type were transferred from the nursery to 32-cm-diameter plastic pots and open pollinated crossing blocks were established. Each hybrid plant was placed in the center and surrounded by three nontransgenic creeping bentgrass plants and three nontransgenic plants of the other *Agrostis* parental species. The crossing blocks were spaced 30 feet apart from each other. The design of the crossing blocks was intended to maximize the chance for crossing. All plants for each crossing block were chosen based on having panicles at the same developmental stage. All of the plants were checked and confirmed to be shedding pollen at the same time.

Seed from all plants were harvested individually and screened in the greenhouse for herbicide-resistant progeny. For screening, all the seed obtained from each plant was sown into 28- by 54-cm flats of commercial potting mix (Pro-Mix, Premier Brands Inc., Red Hill, PA). After 3 to 4 wk of growth, the seedlings were sprayed with Finale as described above. Approximately 2 wk after the spray, a representative sample was counted for survivors and dead plants.

Results

Self Fertility in Agrostis *spp.*

Agrostis species are generally considered to be self incompatible, although there is a low level of selfing. Therefore, most seed produced from a plant originates from fertilization from pollen from another individual. To assess the potential for self-fertility in the *Agrostis* species used in this study, unopened panicles from an individual plant were bagged together. The data obtained regarding self-fertility in the four *Agrostis* spp. Low numbers of plants were recovered for all four species. Self-fertilization can therefore occur in these species, but at a low frequency. The number of florets per panicle can vary considerably, but is generally >100. On the basis of such a conservative estimate of potential seed set, the frequency of selfing observed ranged from 0.2 to 0.5%.

Interspecific Hybridization of Creeping Bentgrass

The results of the crosses between the transgenic creeping bentgrass plants and velvet bentgrass, dryland bentgrass, redtop bentgrass, and colonial bentgrass. Transgenic interspecific hybrids were recovered from all four *Agrostis* spp. Both transgenic lines resulted in interspecific hybrids with all four species, with the exception of dryland bentgrass where no hybrids were recovered when transgenic line 4475 was the pollen parent.

Overall, however, low numbers of progeny were obtained. In all cases, the number of herbicide-resistant progeny was considerably lower than the expected 50%, suggesting that many of the plants obtained from those crosses originated from self fertilization rather than interspecific hybridization. Although herbicide-resistant hybrids were recovered from all four species tested, in some of the individual crosses no hybrids were obtained. This suggests there is variation in capacity for interspecific hybridization with creeping bentgrass among individuals within each of the four related *Agrostis* spp.

The transgenic hybrids were identified as plants that survived two rounds of spraying with glufosinate. To confirm the presence of the *bar* gene, DNA was isolated from randomly selected hybrid individuals and subjected to PCR analysis. The herbicide-resistant phenotype of the hybrids was correlated with the presence of the *bar* gene for all samples tested. No amplification was detected in nontransgenic plants of the four *Agrostis* spp. used

Fertility of the Herbicide-Resistant Interspecific Hybrids

In order for interspecific hybridization to be useful in a breeding program, the primary hybrids must have some level of fertility. Three open-pollinated crossing blocks were established for each hybrid class to assess the potential for fertility of the hybrids. Progeny from the hybrids and from each nontransgenic plant in the crossing blocks were screened for herbicide resistance.

Recovery of progeny from all of the hybrids themselves indicates fertility through the egg. There was, however, considerable plant-to-plant variation in number of progeny recovered. Pollen from both parental species was available to the hybrids so we do not know if there was any preference in pollination of the hybrids by either of the parental species. Progeny from the transgenic hybrids segregated with ratios ranging between 1:1 to 1:6 (herbicide-susceptible: herbicide-resistant). A ratio of 1:1 might be expected if most of the progeny originated from pollination from one of the nontransgenic parental species. It may be that in some of the hybrid plants self-incompatibility has broken down, thus resulting in higher frequencies of herbicide-resistant progeny from selfing. Alternatively, there may be distortions in segregation ratios during meiosis because of an unbalanced complement of chromosomes. Another possibility is that some of the hybrids may have been preferentially pollinated by transgenic pollen from other transgenic hybrids in other crossing blocks (at least 30 feet away), rather than the pollen from the nontransgenic plants adjacent to them (inches away).

The recovery of herbicide-resistant progeny from the nontransgenic plants in the crossing blocks indicated pollination from the central transgenic hybrid plant. As expected, pollination of the nontransgenic plants in the crossing blocks was mainly from other nontransgenic plants of the same species, so most of the progeny harvested from them were not herbicide resistant. There was variation in the ability of the different hybrid types to pollinate their parental species. All four hybrid types were capable of backcrossing to creeping bentgrass. The hybrids between creeping bentgrass and velvet bentgrass or dryland bentgrass did not, however, backcross readily with velvet bentgrass or dryland bentgrass, respectively. The hybrids between creeping bentgrass and colonial bentgrass or redtop bentgrass did backcross readily to both parental species.

Overall, the results indicate all the transgenic hybrids tested had some level of fertility, both through the egg and the pollen, but there was considerable variation in the potential for crossing. In addition to the variation observed in fertility among the hybrid types, there was variation in fertility among the hybrids within each type. The objective of the crossing blocks was to determine if the hybrids had any fertility, not to compare their level of fertility with that of the parental species. Such comparisons, as well as assessments of self-compatibility and levels of backcrossing, will need to be determined from controlled crosses. However, the data presented here show that the *Agrostis* hybrids can be fertile and that they can be backcrossed to the parental species. For future use of interspecific hybridization in *Agrostis* breeding programs, a herbicide-resistance gene will be a useful tool in hybrid identification.

DISCUSSION

The data reported here indicate that interspecific hybridization could be used in *Agrostis* breeding programs. Hybrids between creeping bentgrass and velvet bentgrass, colonial bentgrass, dryland bentgrass, and redtop bentgrass were recovered and were fertile to varying degrees. In our study, flowering times were artificially manipulated in the greenhouse to have plants at the same developmental stages for crossing. Clearly, all four *Agrostis* species tested are biologically capable of hybridizing with creeping bentgrass. The frequency of hybridization was, however, quite low for all four species. Most of the progeny obtained from the crosses performed were actually selfs from the nontransgenic parent rather than interspecific hybrids. Our data indicates that the frequency of interspecific hybridization is similar to or lower than the frequency of selfing. On the basis of isozyme analysis, no selfing was observed in some bentgrass crosses (Warnke et al., 1998). In those cases, there was pollen available for outcrossing, which may have outcompeted any self-pollination. In the study reported here, where there was no pollen available for outcrossing, self pollination and interspecific hybridization occurred at similar frequencies.

In a companion study investigating the potential for interspecific hybridization of transgenic creeping bentgrass under field conditions, a low level of hybrids was recovered from colonial bentgrass and dryland bentgrass (Belanger et al., 2003). No hybrids were recovered with velvet bentgrass or redtop bentgrass. This is certainly because of the fact that, under field conditions, there was no overlap in flowering time between the creeping bentgrass and these two species. Wipff and Fricker (2001) reported a similar low recovery of interspecific hybrids of creeping bentgrass with other *Agrostis* spp. under open-pollinated conditions.

Our data emphasize the importance of a selection method for identification of any hybrids produced. An easily screenable phenotype, such as the herbicide resistance trait used here, makes hybrid identification simple. Other methods could also be used, such as the development of molecular markers specific for either parent. Progeny from interspecific crosses could then be screened by PCR to identify hybrids and eliminate selfs.

Although interspecific hybridization between creeping bentgrass and other *Agrostis* spp. occurs at low frequencies, it may be a useful new method that could be implemented in *Agrostis* breeding programs. Creeping bentgrass is generally quite susceptible to the fungal disease dollar spot caused by *Sclerotinia homoeocarpa* F.T. Bennett, whereas colonial bentgrass has good resistance (Plumley et al., 2000). Conversely, colonial bentgrass generally is quite susceptible to brown patch (*Rhizoctonia solani* Kühn), whereas creeping bentgrass has good resistance. Interspecific hybridization between creeping bentgrass and colonial bentgrass and backcrossing to each parental species may be a way of improving the disease resistance of both species. These two species have the same chromosome number, which may make introgression of the desirable genes from interspecific hybrids feasible. In a field test, some of the hybrids recovered between creeping bentgrass and colonial bentgrass were found to have excellent dollar spot resistance (Belanger, Bonos, and Meyer, 2002, unpublished data). Such dollar-spot-resistant hybrids may offer opportunities for cultivar improvement through gene introgression and for identification of the resistance genes involved

Transgenic crops in agriculture: Possible Risks Associated with Using

Some consumers and environmentalists feel that inadequate effort has been made to understand the dangers in the use of genetic engineering, including their potential long-term impacts. Some consumer-advocate and environmental groups have demanded the abandonment of transgenic crop research and development. Many individuals, when confronted with conflicting and confusing statements about the effect of transgenic crops on our environment and food supply, experience a "dread fear" that inspires great anxiety. This fear can be aroused by only a minimal amount of information or, in some cases, misinformation. With people thus concerned for their health and the well-being of our planetary ecology, the issues related to their concerns need to be addressed. These issues and fears can be divided in to three groups: health, environmental, and social.

Health-related Issues

Allergens and Toxins

People with food allergies have an unusual immune reaction when they are exposed to specific proteins, called allergens, in food. About 2 percent of people across all age groups have a food allergy of some sort. The majority of foods do not cause any allergy in the majority of people. Food-allergic people usually react only to one or a few allergens in one or two specific foods. A major safety concern raised with regard to genetic modification technology is the risk of introducing allergens and toxins into otherwise safe foods. The Food and Drug Administration (FDA) checks to ensure that the levels of naturally occurring allergens in foods made from transgenic crops have not significantly increased above the natural range found in conventional foods. And, transgenic technology is being used to *remove* the allergens from peanuts, one of most serious causes of food allergy.

Antibiotic Resistance

Antibiotic resistance genes are used to identify and trace a trait of interest that has been introduced into plant cells. This technique ensures that a gene transfer during the course of genetic engineering was successful. Use of these markers has raised concerns that new antibiotic-resistant strains of bacteria will emerge. The rise of diseases that are resistant to treatment with common antibiotics is a serious medical concern of genetic engineering opponents.

The potential risk of transfer from plants to bacteria is substantially less than the risk of normal transfer between bacteria, or between us and the bacteria that naturally occur within our alimentary tracts. Nevertheless, to be on the safe side, FDA has advised food developers to avoid using marker genes that encode resistance to clinically important antibiotics.

Environmental and Ecological Issues

Potential Gene Escape and Development of "Superweeds"

There is a belief among some opponents of genetic engineering that the new crops might cross-pollinate with related weeds, possibly resulting in "superweeds" that become more difficult to control. One concern is that pollen transfer from glyphosate-resistant crops to related weeds can confer resistance to glyphosate. While the chance of this happening, though extremely small, is not inconceivable, resistance to a specific herbicide does not mean that the plant is resistant to other herbicides, so affected weeds could still be controlled with other products.

Some people are worried that genetic engineering could conceivably improve a plant's ability to "escape" into the wild and produce ecological imbalances or disasters. Most crop plants have significant limitations in their growth and seed dispersal habits that prevent them from surviving long without constant nurture by humans, and they are thus unlikely to thrive in the wild as weeds.

Impacts on "Nontarget" Species

Some environmentalists maintain that once modified crops have been released into the environment, they could have unforeseen and undesirable effects. Although transgenic crops are rigorously tested before being made commercially available, not every potential impact can be foreseen. *Bt* corn, for instance, produces a very specific pesticide intended to kill only pests that feed on the corn. In 1999, however, researchers at Cornell University found that pollen from *Bt* corn could kill caterpillars of the harmless Monarch butterfly. When they fed Monarch caterpillars milkweed dusted with *Bt* corn pollen in the laboratory, half of the larvae died. But follow-up field studies showed that under real-life conditions Monarch butterfly caterpillars are highly unlikely to come into contact with pollen from *Bt* corn that has drifted onto milkweed leaves — or to eat enough of it to harm them.

Insecticide Resistance

Another concern related to the potential impact of agricultural biotechnology on the environment involves the question of whether insect pests could develop resistance to crop-protection features of transgenic crops. There is fear that large-scale adoption of *Bt* crops will result in rapid build-up of resistance in pest populations. Insects possess a remarkable capacity to adapt to selective pressures, but to date, despite widespread planting of *Bt* crops, no *Bt* tolerance in targeted insect pests has been detected.

Loss of Biodiversity

Many environmentalists, including farmers, are very concerned about the loss of biodiversity in our natural environment. Increased adoption of conventionally bred crops raised similar concerns in the past century, which led to extensive efforts to collect and store seeds of as many varieties as possible of all major crops. These "heritage" collections in the USA and elsewhere are maintained and used by plant breeders. Modern biotechnology has dramatically increased our knowledge of how genes express themselves and highlighted the importance of preserving genetic material, and agricultural biotechnologists also want to make sure that we maintain the pool of genetic diversity of crop plants needed for the future. While transgenic crops help ensure a reliable supply of basic foodstuffs, U.S. markets for specialty crop varieties and locally grown produce appear to be expanding rather than diminishing. Thus the use of genetically modified crops is unlikely to negatively impact biodiversity.

Seeds of Modernisation

The agricultural economy is mainly based on a combination of agro-based and livestock-oriented activities. Agriculture has remained its mainstay (80 per cent of the population lives in rural areas) although traditional business communities have pursued their own interests and new industrial groups are gaining a foothold.

Rajasthan's agrarian landscape is varied. Apart from food crops such as wheat, rice and bajra, fodder crops are also grown. If previously coarse cereals such as bajra and jowar were the predominant crops in the cultivated areas, now there are areas under basmati rice and soyabean as well. It is a little-known fact that the State is one of the leading producers of cumin seeds or *jeera,* a favourite and an essential spice in Indian cuisine. Jalore, known for its granite mining, is equally famous for *jeera* cultivation. Rajasthan leads in the cultivation of gowar bean, moth bean and coriander. Nearly 80 per cent of the gowar produced in the country comes from Rajasthan and the State ranks first in moth bean production. The total area covered under these crops is close to four lakh hectares and it is estimated that four million families depend on them for their livelihood. *Isabgol* or flea husk and *methi* or fenugreek are important medicinal and spice crops grown in Rajasthan. The average area sown during the past five years under *isabgol* was 88,000 ha and the average production was 60,000 tonnes. Similarly, the average area sown under *methi* was 470,000 ha and the average production 54,000 tonnes. The State government now wants the Central government to declare a minimum support price for some of these crops.

The State government plans to establish 15 commodity-specific *mandis* or granaries and develop agricultural export zones and agro-food parks. It also proposes to encourage contract farming in selected areas, enhance rural credit flow to farmers and intervene directly in the market through crop and weather insurance schemes. A State bio-fuel mission has been established with the objective of exploring alternative sources of diesel by promoting the cultivation of multi-purpose crops such as *Jatropha curas,* karanj *(Pongamia pinnata)* and castor and other medicinal plants. The mission also envisages providing land to raise nurseries and plantations and arranging the supply of quality seeds/saplings for plantation, ensuring farmers' access to markets and providing research and development facilities and financial support to set up plantations and oil extraction plants.

A new agricultural policy was announced in 2004. Some 75,000 new tubewell connections were given and a decision was taken to dispose of in three years cases that had been pending for the past 12 years. The Bharatiya Janata Party (BJP) government claims credit for the higher number of connections provided during its tenure compared to the 15,000-20,000 connections given by the previous government. The Indira Gandhi Canal or the Rajasthan Canal irrigates the arid western districts of Bikaner and Jaisalmer. Ganganagar district is irrigated with water from the Ganga Canal. The State government provided Rs.5.53 crores for cleaning the irrigation canals in the tail end of the Indira Gandhi Nahar Project (IGNP) area. Perhaps having learnt its lessons from the recent farmers' agitation, the State government has decided to take a serious look at agrarian issues. For the first time, a sum of Rs.42 crores has been allocated for canal maintenance. An intake structure costing Rs.10 crores has been created at the Indira Gandhi RD 629 (a part of the canal-irrigated area), which will utilise 10,000 cusecs of water from the Ghaggar river. Under drought relief work, construction of canals was undertaken in the irrigated areas of Ganganagar, Hanumangarh, Bikaner and Jaisalmer districts.

Keeping in mind the vagaries of nature and the frequency of drought conditions in the State — as of now 22 of the 32 districts are under a spell of drought — and as part of its policy to provide insurance cover for agriculturists, the government has started an Agriculturist Life Welfare Planning Scheme. As for crop insurance, 14 crops were insured in 2004-05. Claims submitted by farmers during the 2004 kharif season were settled, and the Rs.133.04 crores disbursed thus benefited more than three lakh farmers. For the Rabi crop, eight crops were insured and an amount of Rs.10.27 crores was claimed in 2004-05. For the first time, crops such as coriander and orange were insured. Agricultural call centres were set up to give free advice to farmers. Agriculturists from various parts of the State were trained and exhibitions were organised to display *aswagandha*, *saonamukhi*, white *musli*, *tulsi*, *kalmegh*, *isabgol* and *henna* crops.

The government has increased its budget for agriculture sixfold. As a result of a combination of efforts and a-not-so-dry year, the production of mustard in 2004-05 touched a record high of 38.58 lakh tonnes; groundnut production also remained high at 4.47 lakh tonnes, and cotton production, despite the farmers' unrest in the major cotton producing belt of Ganganagar, is also expected to touch a record 11.24 lakh bales. The government has launched special measures to step up production in various parts of the State. For instance, if it is mustard in Bharatpur and Nagaur, it is jowar in Jaipur and Sikar.

There are plans to create infrastructure and R&D facilities for agriculture. Two new seed laboratories are planned to be set up in Alwar and Jodhpur. Similarly, two pesticide-testing facilities, in Jodhpur and Kota, and two tissue culture laboratories, in Jhalawar and Ganganagar, are on the cards. Twelve soil-testing laboratories are being set up in Baran, Dholpur, Karouli, Churu, Jhunjhunu, Jalore, Hanumangarh, Jaisalmer, Barmer, Rajsamand, Dausa and Boondi districts.

For the first time, a national gardening mission has been launched in 13 districts of the State — Jaipur, Ajmer, Alwar, Chittor, Kota, Baran, Jhalawar, Jodhpur, Pali, Jalore, Barmer, Nagaur and Ganganagar. Under this, Rs.14.50 lakhs has been provided for the development of 16 government and private nurseries.

Overall, there have been some significant interventions in agriculture. Special *mandis* such as the garlic *mandi* in Badoud, the coriander *mandi* at Ramganj, the *jeera mandi* at Bhinmal, the *henna mandi* at Sojat and the fruit *mandi* at Pushkar in Ajmer district have been set up. For the development of the *mandis* it has become imperative to develop roads. Under the Missing Link Project, nearly 826 km of road has been laid with an investment of Rs.73.87 crores in 10 districts. Repair work was done on 2,946 km by investing Rs.84.42 crores.

It is clear that the State government wants to move beyond traditional agriculture. Commercial fruit farming and development of granaries for cash crops are indicative of the government's move not only to diversify agricultural production but also to develop marketing strategies.

Hybridization Between Crops and Wild Plants in the Age of Genetic Engineering

The global debate about genetically modified (GM) crops remains heated and highly polarized. In 2003, almost 68 million hectares of such crops were grown worldwide by an estimated 7 million farmers in 18 countries, an area which continues to grow year by year at a double-digit rate (James, 2003. Most (73%) of the area was occupied by herbicide-tolerant soybean, maize, canola, and cotton. Most of the rest was insect-resistant (Bt) maize and cotton, but the stacking of these traits and the cultivation of crops with other traits such as virus resistance has continued to increase. Eleven of the countries growing GM crops last year were in the developing world where farmers appear increasingly keen to adopt the technology. By contrast, Europe, with minor exception (a small acreage of Bt maize) continues to oppose both the commercial cultivation of GM crops and, as far as trade rules allow, their importation. European consumers demand labeling and segregation of the products of such crops, and, led by highly successful and well-trusted environmentalist nongovernment organizations, the public have consistently opposed their cultivation on the grounds of potential environmental damage. Recently the UK government, displaying what is arguably a bizarre sense of proportionality, has banned transgenic herbicide tolerant sugar beet and canola because management of the crop could exacerbate the decline in farmland biodiversity, especially farmland birds, whilst taking no action either to reduce the widespread growth of winter sown, as opposed to spring, crops (identified as a major factor in the decline of several of the bird species concerned), or to prevent the extensive conversion of pasture to year-on-year forage maize (which has a huge environmental impact).

One aspect of this debate that has created considerable angst has been the problem of gene flow—specifically the "escape" of transgenes into surrounding populations of other crops or of related wild plants. Both the scientific challenges and the potential difficulties for accurate and sensible risk assessment of gene flow from GM crops were recognized very early (Colwell et al., 1985 and the possible spread of novel genes in the environment was highlighted as "an important uncertainty" in risk assessment in the report of the Royal Commission on Environmental Pollution in 1989 (RCEP, 1989 The likelihood of widespread genetic "pollution" and the spectre of "superweeds" created by gene flow has been a central plank in the anti-GM stance of several pressure groups (e.g., Greenpeace; Fronwald and Straus, 1998. A burgeoning scientific research program on the measurement and possible consequences of gene flow (e.g., Lutman, 1999 Den Nijs et al., 2004. has been accompanied by growing investment in research into molecular and other methods of gene containment, or mitigation of the effects of gene escape (Daniell, 2002 Gressel, 2002. Gene flow is a hot topic.

Into this maelstrom arrives this extremely timely volume by Professor Norman Ellstrand of the University of California, Riverside. Professor Ellstrand, more than most, has been responsible for alerting a wider audience than research scientists to the fact and ubiquity of hybridization between crops and their wild relatives. This remains the clearly stated aim of his book. Indeed, it is structured specifically to appeal to a broad audience and the author invites different kinds of reader to approach the book differently, suggesting that evolutionary plant geneticists may like to skip or skim some chapters, whilst practical scientists such as policy makers could jump around from chapter to chapter and yet others take a "linear approach." The overall effect, if like me, you do take a linear approach, is rather curious; the style and language changes abruptly from place to place to address the various audiences. Thus at times we are told that stray pollen grains "have considerably more significance than a stolen kiss" or that beet plants have "children" and "grandchildren," whereas key technical concepts such as "effective population size" or words such as "ribosomal DNA markers" are introduced elsewhere into the text without qualification or explanation. This makes for something of a roller coaster read but, as was no doubt the intention, provides something to suit most tastes. The more waggish mode is used to headline the sections of the book. Part 1 "Foreplay" introduces the subject of hybridization and gene flow and considers some of its evolutionary consequences, part 2 "Caught in the Act" examines the evidence that can be used to detect that hybridization has occurred and catalogs the many examples of crop/wild relative hybridization and possible introgression, and part 3 "Dangerous Liaisons?" deals with the potential impacts of gene flow to wild relatives and the "special" case of genetically engineered plants. Embedded within these main sections is a three-part narrative "The Case of the Bolting Beets," which, under the headlines "Long-Distance Romance," "Detectives Arrive at the Scene," and "The Cloudy Crystal Ball," gradually unfurls the fascinating (and from the GM risk assessment point of view, highly relevant) story of the origins of weed beet in European agriculture via hybridization between wild sea beet *Beta vulgaris* subsp. *maritima* and the domesticated sugar beet *Beta vulgaris* subsp. *vulgaris*. (Incidentally, the playful style is very infectious—I wanted to call this review "Professor Shows Wild Sex is the Norm.")

In a book explicitly designed to have more or less stand-alone bits, which are the best bits? For me, they are the core, which detail what is actually known about crop/ wild relative hybridization, and the ultimately balanced account of the issues presented by gene flow from GM crops.

Many people will recognize the core chapters as an extension of an earlier review (Ellstrand et al., 1999) of the known cases of spontaneous hybridization between the most widely grown crop plants and their wild relatives. This is the most original part of the book and covers a field to which Professor Ellstrand and his students have made a unique and extensive contribution. If there is anyone left out there who believes that hybridization between domesticated and wild plants is an exceptional event they should read this thoroughly researched account covering most of the world's crops. In all but three of the 25 main crops looked at in detail, there is some evidence for natural hybridization somewhere in the world with one or more relatives, and in all but seven of the 28 species represented by those crops, there is modern molecular evidence to support the existence of hybrids. A fuller list of 81 cultigens in which there is evidence of crop/wild relative hybridization contains only 33 examples in which hybridity has been deduced from morphological evidence alone. However, evidence of the occurrence of hybrids is not the same thing as saying that gene flow, which includes introgression and the incorporation of a gene into a natural population, is a ubiquitous feature of contemporary agriculture—a subject to which I will return later. The specific problems presented by GM crops (Ch. 11), in which the traits inserted by recombinant DNA methods are almost unexceptionally inherited in a dominant mode (which, it is important to note, exposes them more rapidly to selection), and discuss the possibilities of containing or limiting gene flow into wild populations, either by methods such as physical separation and buffer zones (rightly concluded to be costly and largely ineffective in most cases) or the sorts of genetic engineering technologies alluded to earlier (which actually bring additional risk assessment issues).

What about the other bits? In a relatively slim volume (just over 200 pages of text), one cannot include everything of relevance, and what goes in will always be a matter of choice. Professor Ellstrand's decision to focus on one-way hybridization, (i.e., where, at least initially, genes are transferred *from* the cultivated plant *to* the wild relative) and also largely to ignore natural ecosystems, obviously makes sense in the context of a discussion of transgene escape from GM crops. However it leaves out research areas that undoubtedly have something to offer in addressing the book's central question. For example, missing from the introductory chapters is a sense of the antiquity of the subject of spontaneous crop/related plant hybridization; needless to say, the topic did not begin with the GM crop debate. Even Charles Darwin (remarkably not cited) refers to the serious contamination of his cabbage seed stocks by kale plants growing half a mile away (Darwin, 1876). It was actually surprising to be told that the prevailing view in the late 1980s was that hybridization between crops and their wild relatives occurred infrequently, if at all. Extensive true gene flow (sensu Futuyma [1998] in which a gene is successfully incorporated from one population into another) may have been thought to be rare in such circumstances (and, as it turns out, probably still is), but the frequent occurrence of crop/wild hybrids was surely acknowledged by many (see e.g., Stace, 1975. Such hybridization was certainly described in the literature on introgression, natural gene exchange between wild relatives and crops in the reverse direction, the evolution of crop mimics, and of crop–weed–wild complexes. This literature, given the one-way focus of the book, has much less prominence than it probably deserves. Similarly, the extensive work in the fields of ecological and conservation genetics on the genetic structure of natural plant populations and the interplay between gene flow and natural selection is barely included. Surely it is these literatures that we are going to have to research both for understanding and for relevant models to predict whether or not, in specific cases, gene flow from GM crops is likely to have a negative impact on the environment. The work on the evolution of herbicide tolerant weeds provides a striking example.

From the viewpoint of assessing the potential risks of cultivating GM crops, the central question posed by the book's title (are the liaisons between crops and wild relatives in some sense "dangerous"?) is correctly identified as comprising two specific questions. Will hybridization between GM crops and wild relatives lead to the creation of weedy or invasive species (or, more specifically, the evolution of increased weediness in existing wild relatives)? Will hybridization between GM crops and wild relatives increase the risk of extinction of any wild taxa? These are really reverse sides of the same ecological genetic question and are critically dependent on the changes in plant fitness that result from hybridization and subsequent introgression. Dealing with the second question first; we are given a table containing nine examples of hybridization with a cultivated species that appears to be endangering one or more wild populations. Although most of the examples provide data that shows that hybridization has occurred, in only one case, that of natural hybridization of cultivated rice *Oryza sativa* with an endemic wild rice *Oryza rufipogon* subsp. *formosana* in Taiwan, is there evidence of decreased fitness in the wild populations. There is a similar problem with judging how much extinction of wild populations may have occurred as a result of hybridization with a domestic species, although "the overall impression of experts in the field is that it is not rare" (p. 155). The aúthor, who suggests that extinction may be very rapid, hence the lack of examples, observes that the topic "has enjoyed more speculation than data collection" (lovely phrase!) (p. 155) and points out that there is a great deal of relevant research that can be done in this area.

The first question (Will gene flow from GM crops produce weedier or more invasive plants?) is in many ways the key question that the title of this book suggests it has set out to address. It is certainly number one among the frequently asked questions about gene flow and GM crops, a central concern of all risk assessments performed by regulators, and has been repeatedly invoked by environmentalists as a reason to refuse consent to release particular crops. In view of this, it will come as a huge disappointment to many readers to realize that we actually know almost nothing about the consequences of gene flow from crops—GM or conventional. Having been convinced that it happens, and on a larger scale than envisioned heretofore, we arrive at the crunch "so what?" question, only to discover a virtually data-free zone. To be fair, Professor Ellstrand does set out some examples that hint at the generalization that gene flow from domesticates tends to result in an increase of genetic diversity in the wild, and offers some possible explanations for this. And there is a smallish body of research literature that allows us to make some intelligent guesses about the really vital risk assessment question of invasiveness. In fact, Professor Ellstrand goes further and lists 13 (of 16) cases in which hybridization between a cultivated plant and a wild relative have "preceded the evolution of increased aggressiveness of the wild taxon" (p. 152) (Five of these involved the evolution of new taxa and eight were examples of adaptive evolution.) He goes on to conclude from this that hybridization between wild plants and their domesticated relatives "has resulted in some plants that have created hardship for humankind and others that have disrupted natural ecosystems"(p. 154) and, by comparison with a data set (of 28 examples) not published in the book, that "the chances of a domesticate-wild hybrid becoming a problem plant are an order of magnitude greater than if both hybridizing parents are wild species"(p. 154). This surprising, and for some readers no doubt alarming, conclusion becomes less surprising if the data are reexamined in the light of the author's definition of "wild" plants, given (on p. 10) as "those that are essentially free-living, that is, those that grow and reproduce without being deliberately planted and cared for." This categorization clearly includes agricultural weeds and the species commonly found in peri-agricultural and frequently disturbed biotopes. And sure enough, the 13 examples of increased aggressiveness include no less than nine weeds of agricultural fields, and of the other four at least two, *Rhododendron ponticum* and *Nasturtium sterile*, are arguably inappropriate analogs for cases of gene flow from domestic crops (their success as invaders can also be given more parsimonious non-genetic "explanations" involving the exploitation of vacant niches).

The distinction between agricultural or disturbed-land weeds and plant species invasive of natural or semi-natural habitats is an extremely important one from the point of view of ecological risk assessment of GM crops. But it is one that is rarely made explicit. A potential weed is a potential agricultural problem and the risk of one resulting from gene flow, or increased weediness of the GM plant, must be set in the context of potential agricultural solutions. Weed management and control strategies have developed alongside evolving weed problems to variously cope with the threats to crop yield. Thus, an assessment of the risks of, say, creating a herbicide-tolerant weed as a result of gene flow from the crop must consider what strategies are available for risk mitigation or management, such as alternative herbicides or changes in farm practice, and decide what is relatively "safe" in that case; no agriculture is entirely risk free. (In fact, such a weed has been reported from the canola fields of Canada (Warwick et al., 2003, probably since the text of *Dangerous Liaisons* was completed, adding to the 130 or so herbicide-tolerant weeds produced by the selective action of herbicides on weed populations.) By contrast, the creation by gene flow of a hybrid that becomes more invasive of natural ecosystems is a very different prospect (and could in theory be a potential nightmare) and is one of the major reasons why GM crops receive regulatory oversight and why they should be scrutinized on a case-by-case basis. One is bound to say, however, that Professor Ellstrand's data actually lend more support to those who maintain that, as a general rule, hybrids that receive genes from modern crops are likely to be less fit in natural habitats because they are invariably linked to genes expressing traits selected during domestication, such as seed shattering and lack of dormancy(e.g., Conner et al., 2003 ; Stewart et al., 2003. It seems that, despite widespread crop/wild plant hybridization there have been virtually no "escapes." Until we have more case studies(and not ones that use fecundity as a substitute for fitness but undertake actual life history analysis—see Bullock, 1999 and Crawley et al., 2001), data such as those provided in this volume, whilst confirming that mechanisms exist for the evolution of more serious crop weeds, give limited clues to the potential for invasiveness and disturbance of natural ecosystems.

Despite this, I feel sure that the book will be widely acclaimed and play an important part in the debate about GM crops. Regulators, policy makers, and the scientists who advise politicians (described by the author as being "in the trenches") will find the extensive information on hybridization in the world's crops and their wild relatives useful and through it gain access to other studies relevant to their own countries. Those scientists in the European trenches, where the debate is rarely leavened by humor and where the distinction between what is dangerous, undesirable, or plain daft has become almost irreversibly blurred, will no doubt have winced at the book's title, even with the question mark, but should read it nonetheless. Students of evolution and plant genetics will find the introduction to the subject informative and clear. The text is rich in quotable phrases (which one predicts will be selected by both sides in the debate as appropriate) and, as indicated earlier, the book can be read at several levels.

Finally, what about the case of the bolting beets? Is it an exception that proves the rule? Whilst it certainly shows what can happen when cultivated plants mate with their wild relatives, unfortunately for Professor Ellstrand's line of argument, weed beets were created by gene flow *into* the seed crops *from* the annual wild beet (sea beet is an annual in the southern part of its range), and hybrid seed were then transported with the crop to the sugar beet production areas in the north where they created an agricultural weed problem. Although the mechanism exists for this type of gene transfer to happen with a transgenic crop, it is hardly a good analogy for gene flow from a GM crop into populations of a wild relative—where the hybrids must survive in the habitat of the wild plant. To me, the most illuminating aspect of the bolting beet story is how little gene flow, if any, has occurred in the north between the crop or weed and the wild sea beet relative. In recent studies using molecular markers, by Joel Cuguen and his colleagues at the University of Lille, populations of sea beet on the north French coast displayed no evidence of gene flow by pollination with beet in nearby fields, but, intriguingly, from the distribution of the cytoplasmic male sterility marker gene, demonstrated clear evidence of seed-mediated gene flow into these populations (the likely vector being tractor tires or sea wall engineers) (Arnaud et al., 2003). Similarly, and despite decades of exposure to pollen from coastal sugar beet fields containing weed beets, the bolting gene has not been found in any wild sea beet populations along the French channel coast or the coast of England (Van Dijk et al., 1997) Sea beet is a perennial plant in these northern sea wall and salt marsh populations. Suggesting that annual beets are less fit in such habitats, this example surely provides yet further testimony to the power and ubiquity of natural selection.

3

Agricultural Revolution & Modern Agricultural Practices

The Agricultural Revolution

The Agricultural Revolution occurred between 1750 and 1900 when the way in which farmers produced food in this country changed. In 1750 most people lived and worked in villages producing food. As the country went through the Industrial Revolution though it was necessary to increase the amount of food grown. This was because the population was increasing quite rapidly.

In the early part of the eighteenth century most farmers had strips of land that they would grow their food on. This system had many disadvantages. Banks of earth separating the strips were wasted land, drainage was poor and because the farmers new little about fertilisers they had to leave land fallow (unused) every four years.

It can be argued that there was no Agricultural Revolution as the changes in farming were gradual. By the early nineteenth century farmers were beginning to use the Norfolk Crop rotation system. This system meant that no land had to remain fallow. The system worked like this:

Each area of land would be split into four sections. The crop that was grown on each field would be rotated so that different nutrients would be taken from the land. In the first year turnips or another root crop would be grown; in the second year barley was grown in the field (barley could be sold at a profit); in the third year clover or a grass crop was grown and in the fourth year wheat was grown in the field (wheat could also be sold for a profit).

As the demand for food increased people began to make improvements to the types of machines used on farms. A seed drill and threshing machine being amongst the first new inventions to help farmers. Inventions such as these, along with the enclosure of fields, helped agriculture to develop rapidly and produce enough food for the growing population.

Agricultural Science in Kautilya's *Arthasastra*

In a recent paper, Y L Nene has discussed the concepts of agriculture of the times of Kautilya, the famous Chanakya who wrote the *Arthasastra* in the 3rd – 4th century BC. The name of Chanakya is much revered in Indian history as he is credited to have put Chandragupta, a common man at that time, on the throne of Magadh through his sheer intelligence and practicality. The treatise, *Arthasastra* written by this wise teacher Kautilya deals with his own views on managing and increasing resources of an empire for its own benefit and consolidation.

Since the objective of Nene is to draw a likeness between the agricultural practices in the past and those in the modern times he points out that agriculture of today receives policy and administrative support from government officials. For production of crops, supply of good seed and other purchases inputs are arranged. Assistance is provided in making available other resources such as labor, machinery, implements, and bullocks or tractor power. The India Meteorology Department makes a general prediction of rainfall for every monsoon. Efforts to predict yields using crop models are made. Contingency plans are made for alternative crops in case the monsoon fails or floods occur. Irrigation is provided wherever the water source exists. Arrangements are made to protect crops and to harvest market, and safely store the produce.

The main points that are brought forth in these paragraphs relate to agriculture and the paragraphs from 4 to 20 (with the exception of 9, 13 & 14) have direct relevance to the general farmers. Main points so covered can conveniently be put as managing agriculture, measurement of rainfall, astrology and astronomy related to agriculture, meteorological aspects of agriculture, kinds of crops, advantage in cultivating certain classes of crops, the growth of crops, treatment of seeds, manuring and harvesting. We can say that the study includes all the vital aspects of agriculture, as we practice it today.

> "Possessed with knowledge of the science of agriculture, water management, and managing crops and trees, or assisted by those who are trained in such sciences, the superintendent of agriculture shall in time collect the seeds of all kinds of grains, flowers, fruits, vegetables, bulbous roots, roots, fruits of creepers, fiber-producing plants such as hibiscus and cotton."

In this connection, paragraph 3 is very relevant to the state agriculture departments. It sounds like a clear order to the superintendent to make sure that the farmers and the laborers are not hampered in their activities and that their work does not suffer. It reads,

> "The work of the above men shall not suffer on account of any want in ploughs and other necessary implements or of bullocks. Nor shall there be any delay in procuring to them the assistance of blacksmiths, carpenters, basket sellers, rope-makers, as well as those who catch snakes, and similar persons. Any loss in production because of the above persons should invite fine equal to the loss."

It is clear that the passage has direct relevance to the agriculture departments as still the farmers depend on them to a large extent for the agriculture activities and which the departments' negligence may affect.

Paragraphs 4, 5 & 8 specifically refer to the meteorological aspects of agriculture and the measurement of rainfall. They show the high level of meteorological knowledge the people possessed in those times. Quite interesting is the 8th one, which tells us about the different clouds bearing the rains, as do the meteorologists of our own day. It says,

> "Three are the clouds that continuously rain for seven days; eighty are they that pour minute drops; and sixty are they that appear with the sunshine – this is considered good, well distributed rainfall. Where rains interspersed with wind and sunshine is such that cow dung cakes can be dried three times (during the rainy season), reaping of a good harvest is certain."

Paragraphs 6 & 7 relate to the astrological/astronomical aspects. These aspects are also dealt with in another ancient text, the *Krishi Parashar*, the original version of which is supposed to be of an earlier date than the *Arthasastra*. Nene has elsewhere pointed out that the astrological aspects of agriculture in the time of sage Parashar had some shortcomings and the search for a better system culminated in the *Arthasastra* of Kautilya.

The paragraph 15 as taken by Nene is interesting in that it tries to show the advantage in cultivating different classes of crops, which is true even today. It reads,

> "Rice crops and the like are the best. Vegetables are intermediate; and sugarcane is the worst (very difficult to grow) for it is subject to various evils and requires much attention and expenditure to reap."

This paragraph is quite relevant to the general community of the farmers as well as the state agricultural departments as they can advantageously use the advice to their benefit.

So we find that in the text of *Arthasastra*, as shown by Nene, there are many interesting points that are worth taking note of and the validity of quite a few can be tested in order to benefit the farmers and the world at large today. While talking of sustainable farming the following paragraph might be relevant,

"The seeds of grains are to be exposed to mist and heat for seven nights; the seeds of kosi (mung bean, black gram, etc.) are treated similarly for three to five nights; the seeds of sugarcane and the like are plastered at the cut end with the mixture of honey, lard, ghee and cow dung; seeds of bulbous roots with honey and ghee; cotton and hard seeds with cow dung; and pits for trees are to be burnt, and manured with bones and dung of cows on proper occasions."

Thus Kautilya's *Arthasastra* seems to be a treasure-house of scientific information on a variety of subjects which are relevant even to the present times.

Neolithic Revolution

The Neolithic Revolution is the term for the first agricultural revolution, describing the transition from hunting and gathering to agriculture, as first adopted by various independent prehistoric human societies, in various locations. The term refers to both the general time period over which these initial developments took place and the subsequent changes to Neolithic human societies which either resulted from, or are associated with, the adoption of early farming techniques and crop cultivation. The first agricultural revolution spurred major social changes, including a high population density, the organization of a hierarchical society, specialization in non-agricultural crafts, a standing army, barter and trade, and the expansion of man's "control" over nature.

The hunter-gatherer way of life was replaced by domestication of crops and animals, enabling people to live more sedentary lives. Permanent settlements arose, creating new social, cultural, economic and political institutions. In this period, agriculture was subsistence agriculture: people farmed for their own subsistence (not for sale or profit), and farmers practiced crop rotation (letting fields lie fallow between planting seasons).

The Neolithic Revolution is notable primarily for developments in social organization and technology. The changes most often associated with the Neolithic Revolution include an increased tendency to live in permanent or semi-permanent settlements, a corresponding reduction in nomadic lifestyles, the concept of land ownership, modifications to the natural environment, the ability to sustain higher population densities, an increased reliance on vegetable and cereal foods in the total diet, alterations to social hierarchies, nascent "trading economies" using surplus production from increasing crop yields, and the development of new technologies. The relationship of these characteristics to the onset of agriculture, to each other, their sequence and even whether some of these changes are supported by the available evidence remains the subject of much academic debate, and seems to vary from place to place.

Agricultural Transition

The term *Neolithic Revolution* was first coined in the 1920s by Vere Gordon Childe to describe the first in a series of agricultural revolutions to have occurred in Middle Eastern history. This period is described as a "revolution" to denote its importance, and the great significance and degree of change brought about to the communities in which these practices were gradually adopted and refined.

This involved a gradual transition from a hunter-gatherer mode of subsistence which was practiced by all earlier human societies, to one based more upon the deliberate nurturing and cultivation of crops for the purpose of food production. Evidence for the first beginnings of this process obtained from different regions dated from approximately 25,000 years ago in Melanesia to 2,500 BC in Sub-Saharan Africa, with some considering the events of 9000-7000 BC in the Fertile Crescent to be the most important. This transition everywhere seems associated with a change from a largely nomadic hunter-gatherer lifestyle to a more settled, agrarian-based one, with the onset of the domestication of plants and of a number of animals. The chronology, social foundations, plant genetics, plant morphology and selective mechanisms of humans, and the processes of the spread of agriculture have been documented by archaeologists in many parts of the world where agriculture first arose (e.g. Melanesia, the "Fertile Crescent", Mesoamerica, South Asia, India (the Ganges), Southeast Asia, Peru, the Mississippi, and the Yuan in China).

Incentive to Settle

Nomadic lifestyles are the product of the depletion of the biological potential of a specific location, either through localised overhunting or over gathering, and leads to a movement to a new area where game and foodstuffs are not depleted, allowing the earlier ranges to recover. If sufficient foodstuffs can be gathered on a permanent basis from a specific locality, there is little incentive to move and permanent settlement may result. This will happen whenever local biological productivity is sufficient to permit permanent settlement. Prehistorically it appears to have happened first with certain beachcomber or lacustrine cultures, such as the Jomon of Japan, the areas of Sundaland and Sahulland of "Greater Melanesia", and historically in the salmon country of the Pacific Northwest region of North America. Natufian sedentism before the Younger Dryas was proto-agricultural in nature, principally around gathering, but not planting crops.

Having a plentiful supply of basic food does not mean that depletion of important gathered vegetable products does not occur. But a settled population permits year-round observation of the growing cycle, and hunter-gatherers are keen observers of the environmental conditions optimal for specific plant products. Research in Australia, with Aboriginal groups in the Australian Northern Territory has shown that they were fully conversant with the biological facts of agriculture even though they never farmed, although, when found, may have been transferred to a more favourable location. Research has shown that only certain crops make good cultivars. In fact most cultivated crops were discovered more than 2,500 years ago, and despite the scientific and technological revolutions, only a few marginal nut crops (e.g. macadamia nuts) have been added in recent times.

TABLE 1

The Worlds's 30 Leading Crops in Terms of Estimated Edible Dry Matter

Food Crop	*Production (mill. tonnes)*	*Plant Type*	*Original Ecosystem*	*Pollination and Polyploidy*
Wheat	468	Annual	Mediterranean	Self 2,4,6
Maize	429	Annual	Savanna	Cross 2
Rice	330	Annual	Savanna, Tropical forest	Self 2
Barley	160	Annual	Mediterranean	Self 2
Soybean	88	Annual	Woodlands	Self 4
Cane Sugar	67	Perennial	Tropical forest	C (veg. prop) many
Sorghum	60	Annual	Savanna, Steppe	Self 2
Potato	54	Annual	Highlands	C (veg. prop) 2,4,6
Oats	43	Annual	Mediterranean	Self 2,4,6
Cassava	41	Perennial	Savanna, Tropical Forest	? 4
Sweet Potato	35	Annual	Savanna	C (veg. prop) 6
Beet Sugar	34	Annual	Coastal	Cross 2,a,4
Rye	29	Annual	Mediterranean, Steppe	Cross 2
Millets	26	Annual	Savanna, Steppe	S/C 2,4
Rapeseed	19	Annual	Mediterranean	Cross 4,6
Beans	14	Annual	Savanna	Self 2
Peanut	13	Annual	Savanna	Self 4
Pea	12	Annual	Mediterranean	Self 2
Musa (bananas, plantain)	11	Perennial	Tropical forest	Vegetative 3
Grape	11	Perennial	Woodlands	- 2
Sunflower	9.7	Annual	Prairie	Cross 2
Yams	6.3	Annual	Tropical forest, Savanna	- many
Apple	5.5	Perennial	Woodlands	C (veg. prop) 2
Coconut	5.3	Perennial	Coastal	Cross 2

Food Crop	*Production (mill. tonnes)*	*Plant Type*	*Original Ecosystem*	*Pollination and Polyploidy*
Cottonseed (oil)	4.8	Annual	Savanna	Cross 4
Oranges, citrus fruits	4.4	Perennial	Tropical forest	C (veg. prop) 2,3
Tomato	3.3	Annual	Coastal	Self 2
Cabbage	3.0	Annual	Coastal	Cross 2
Onion	2.6	Annual	Mediterranean	Cross 2
Mango	1.8	Perennial	Tropical forest	C (veg. prop) 2

Zones of Crop Domestication

The Russian agronomist Nikolai Vavilov suggested that the wild ancestors of these and other domesticated crops were found only in different regions of the world.

1. Mediterranean climate areas of the Middle East — Wheat, barley, oats, rye, grapes, apples, onions, lentils, vetch, garlic, chickpeas, dates, fig, pomegranate, pistachio
2. Loess region of Northern China — Japanese millet, buckwheat, soyabean, apricot, peach, plum, cherry, apple
3. Monsoon forests from Southern China to Bangladesh — Rice, oranges & citrus, pigeon pea, yams, water chestnut, star fruit, durian, lychee
4. Tropical rain forests of New Guinea and adjacent regions — Bread-fruit, banana, plantain, mango, sago, coconuts, taro, sugar cane, pit-pit
5. Ethiopia, the Sudan and the Sahel — Teff, sesame, ensete, sorghum, coffee, pearl millet, Guinea millet, African rice, okra, watermelons, oil palm, cottonseed
6. Mesoamerica (Mexico and Central America) Maize, beans, capsicum, chili, tomatoes, cocoa, squash, sweet potato
7. Andes (Ecuador to Northern Chile) — Potato, cotton and peanuts
8. Central America and Amazonia — Amaranth, cassava, pineapple
9. Mississippi Prairie — Sunflower seeds

Carl Sauer (1952), a geographer whose book "Agricultural Origins and Dispersals" has become a classic, combined the Darwinian views with Eduard Hahn's idea (1896, 1909) that vegetative propagation (crops having lost capacity for seed production, or growing in an environment where seeds were not needed) should precede grain agriculture, and set out to locate the cradle of agriculture on theoretical grounds. He listed 6 presuppositions as a basis for his search:

1. Agriculture did not originate from a growing or chronic shortage of food. People living in the shadow of famine do not have the means or time to undertake the slow and leisurely steps out of which a better and different food supply is to develop in a somewhat distant future.
2. The hearths of domestication are to be sought in areas of marked diversity of plants and animals... This implies well-diversified terrain and perhaps also variety of local micro climates.
3. Primitive cultivators could not establish themselves in large river valleys subject to lengthy floods and requiring protective dams, drainage, or irrigation.
4. Agriculture would begin in wooded but not heavily forested lands. Primitive cultivators could readily use open spaces for planting by deadening trees; they could not dig in sod or eradicate vigorous stoloniferous grasses...
5. The inventors of agriculture had previously acquired special skills in other directions that predisposed them to agricultural experimentation.
6. Above all, the founders of agriculture were sedentary folk.

The sedentary life, he thought, should therefore precede agriculture and could best be developed by fishing tribes, and for his purpose he sought them on fresh waters in a warm climate. Fresh water was selected because seaside vegetation has contributed relatively little to agriculture and what has been developed has come late in crop evolution. With these presuppositions in mind, he proposed that the monsoon forests of Southeast Asia would be the oldest hearth of agriculture. Adopting a diffusionist approach, he suggested that from there agriculture spread to the Indian subcontinent, and Southern China and then to the Middle East and Europe. He proposed an independent origin for American crops, but left open the possibility for influences from across the Pacific from South East Asia.

Archaeology, concentrated in Europe and the Middle East suggested that Middle Eastern farming took place significantly later to that of Europe, where farming tended to be largely if not exclusively, of grain crops and plants originating in the wetter Mediterranean climate and mountain foothill areas of the Middle East. As a result a number of scholars proposed that around 12,500 years ago as the world's climates were changing, hunter-gatherers were forced to turn to alternative methods of obtaining food. Climatic changes over time forced some people to work much harder and travel longer distances in search of food. Over thousands of years, hunter-gatherers unconsciously adjusted to their surroundings. Hunter-gatherers began to stay near reliable sources of water and bring wild seeds back to their base camp to plant nearby. The Australian archaeologist Vere Gordon Childe maintained that the key factor in this change was that global climates at the end of the last ice age became warmer and drier, making plants more efficient at producing crops but encouraging settlement near water sources. But paleoclimatology and the study of sub-fossil pollen demonstrated that after the ice age climates in the region had actually turned wetter, requiring that the forces governing Childe's "Neolithic Revolution" be revised.

The primacy of agriculture in the Middle East was challenged, however, by the findings of French archaeologists in Vietnam, at the site of Hoa Binh. The Hoabinhian culture seemed to confirm Sauer's thesis. By 12,000 BC, the pre-historic people of Vietnam had abandoned its nomadic life to settle in the Hong (Red) River valley. They lived in caves and rock shelters close to water streams and were making ground oval, circular or triangular adzes with a sharp edge. At the Spirit Cave in Thailand Chester Gorman in 1965 found, the Hoabinhians were mostly hunters but they also cultivated a range of plant crops; including almond, water chestnut, broad beans, peas, cucumbers, betel nut, bottle gourd and butternut pumpkin or squash to gather their fruits, nuts or edible roots. Dated between 9000 and 6000 BC by Carbon 14 dating, this figure was later revised upwards by 3,000 years by Wilhem Solheim II of the University of Hawaii, who found cruder forms of the Hoabinh technology from 12,000 BC. Solheim claimed that the inhabitants at Spirit Cave had 'an advanced knowledge of horticulture', and his chronological chart suggests that 'incipient agriculture' began at about 20,000 BC in southeast Asia. This fact suggested to some that domestic cultivation may have existed in South East Asia earlier than in the Near East (Iraq) as many Western historians had until then believed. The finding of pottery, and polished stone tools from 6,600 BC also upset the theory that these technologies only arrived in the region about 3,000 BC from China, and suggested that perhaps the diffusion in fact went in the other direction. The Haobinhian culture also produced a waisted adze, hafted transversely, the first evidence of a hoe confirmed in the world.

Emergence of Civilization

Without agriculture, the emergence of many of the traits popularly referred to as "civilization" would not have been possible (*e.g.* cities, advanced technology, social hierarchies, organized warfare, etc.). The documentation and interpretation of the natural and social changes associated with the origins of agriculture is one of the great success stories of archaeology (particularly environmental archaeology).

Domestication of Plants

Once agriculture started gaining momentum, humans were unknowingly altering the genetic make-up of certain cereal grasses (beginning with emmer, einkorn and barley), and not simply those that would favour greater caloric returns through larger seeds. Plants that possessed traits such as small seeds, or bitter taste would have been seen as undesirable. Plants that rapidly shed their seeds on maturity tended not to be gathered at harvest, thus not stored and not seeded the following season; years of harvesting selected for strains that retained their edible seeds longer. Several plant species, the "pioneer crops" or Neolithic founder crops, were the earliest plants successfully manipulated by humans. Some of these pioneering attempts failed at first and crops were abandoned, sometimes to be taken up again and successfully domesticated thousands of years later: rye, tried and abandoned in Neolithic Anatolia, made its way to Europe as weed seeds and was successfully domesticated in Europe, thousands of years after the earliest agriculture. Wild lentils present a different challenge that needed to be overcome: most of the wild seeds do not germinate in the first year; the first evidence of lentil domestication, breaking dormancy in their first year, was found in the early Neolithic at Jerf el-Ahmar, (in modern Syria), and quickly spread south to the Netiv Hagdud site in the Jordan Valley. This process of domestication allowed the founder crops to adapt and eventually become larger, more easily harvested, more dependable in storage and more useful to the human population.

Barley and, most likely, oats, were cultivated in the Jordan Valley, represented by the early Neolithic site of Gilgal, where in 2006 archaeologists found caches of seeds of each in quantities too large to be accounted for even by intensive gathering, at strata dateable c. 11000 years ago. Some of the plants tried and then abandoned during the Neolithic period in the Ancient Near East, at sites like Gilgal, were later successfully domesticated in other parts of the world.

Once early farmers perfected their agricultural techniques, their crops would yield surpluses which needed storage. Hunter gatherers could not easily store anything as they were on the move constantly, whereas those with a sedentary dwelling could store their surplus grain. Eventually granaries were developed that allowed villages to store their seeds for longer periods of time. So with more food, the population expanded and communities developed specialized workers and more advanced tools.

The process was not as linear as was once thought, but a more complicated effort, which was undertaken by different human populations in different regions.

Agriculture in Asia

The Neolithic Revolution is believed to have become widespread in southwest Asia around 8000 BC–7000 BC, though earlier individual sites have been identified. Although archaeological evidence provides scant evidence as to which of the genders performed what task in Neolithic cultures, by comparison with historical and contemporary hunter-gatherer communities it is generally supposed that hunting was typically performed by the men, whereas women had a more significant role in the gathering. By extension, it may be theorised that women were largely responsible for the observations and initial activities which began the Neolithic Revolution, insofar as the gradual selection and refinement of edible plant species was concerned.

The precise nature of these initial observations and (later) purposeful activities which would give rise to the changes in subsistence methods brought about by the Neolithic Revolution are not known; specific evidence is lacking. However, several reasonable speculations have been put forward; for example, it might be expected that the common practice of discarding food refuse in middens would result in the regrowth of plants from the discarded seeds in the (fertilizer-enriched) soils. In all likelihood, there were a number of factors which contributed to the early onset of agriculture in Neolithic human societies.

Agriculture in the Fertile Crescent

Agriculture first arose in the Fertile Crescent because of many factors. The Mediterranean climate has a long, dry season with a short period of rain, which made it suitable for small plants with large seeds, like wheat and barley. These were the most suitable for domestication because of the ease of harvest and storage and the wide availability. In addition, the domesticated plants had especially high protein content. The Fertile Crescent had a large area of varied geographical settings and altitudes. The variety given made agriculture more profitable for former hunter-gatherers. Other areas with a similar climate were less suitable for agriculture because of the lack of geographic variation within the region and the lack of availability of plants for domestication.

Agriculture in Africa

The Revolution developed independently in different parts of the world, not just in the Fertile Crescent. On the African continent, three areas have been identified as independently developing agriculture: the Ethiopian highlands, the Nile River Valley and West Africa.

Prof. Fred Wendorf and Dr. Romuald Schild, of the Department of Anthropology at Southern Methodist University, have evidence of early agriculture in Upper Paleolithic times at Wadi Kubbaniya, on the Kom Ombos plateau, of Egypt, including a mortar and pestle, grinding stones, several harvesting implements and charred wheat and barley grains — which may have been introduced from outside the region. Carbon-14 dates range from 15,000 to 16,300 BC, showing that this early grain harvesting exceeded that of the Middle East by about 5,000 years.

The archaeologists state that "These are not the only Late Paleolithic sites which have been discovered in Egypt along the Nile, nor or they alone in containing stone artifact assemblages which seem to indicate the harvesting of grain. Among others are several sites at Wadi Tushka, near Abu Simbel, at Kom Ombo, north of Aswan, and a third group [a whole series of sites] near Esna. All these are in the Nile Valley." The Egyptian Esna culture shows "extensive use of cereals," date from 13,000 to 14,500 years ago.

They continue: "While the flaked stone industries from them are different from those found at Kubbaniya, the Tushka site yielded several pieces of stone with lustrous edges, indicating that they were used as sickles in harvesting grain."

Many such grinding stones are found with the early Egyptian Sebilian and Mechian cultures dating 10,000-13,000 BC. Smith writes: "With the benefit of hindsight we can now see that many Late Paleolithic peoples in the Old World were poised on the brink of plant cultivation and animal husbandry as an alternative to the hunter-gatherer's way of life". Unlike the Middle East, this evidence appears as a "false dawn" to agriculture, as the sites were later abandoned, and permanent farming then was delayed until 4,500 BC with the Tasian and Badarian cultures and the arrival of crops and animals from the Near East.

Domestication of Animals

When hunter-gathering began to be replaced by sedentary food production it became more profitable to keep animals close at hand. Therefore, it became necessary to bring animals permanently to their settlements. The animals' size, temperament, diet, mating patterns, and life span were factors in the desire and success in domesticating animals. Animals that provided milk, such as cows and goats, offered a source of protein that was renewable and therefore quite valuable. The animal's ability as a worker (for example ploughing or towing), as well as a food source, also had to be taken into account. Besides being a direct source of food, certain animals could provide leather, wool, hides, and fertilizer. Some of the earliest domesticated animals included sheep, goats, cows, and pigs. Out of the thousands of species of animals only fourteen eventually became domesticated for agricultural purposes.

Domestication of Animals in the Middle East

The Middle East served as the source for many domesticable animals, such as goats and pigs. This area was also the first region to domesticate the Dromedary Camel. The presence of these animals gave the region a large advantage in cultural and economic development. As the climate in the Middle East changed, and became drier, many of the farmers were forced to leave, taking their domesticated animals with them. It was this massive emigration from the Middle East that would later help distribute these animals to the rest of Afroeurasia.

Domestication of Animals in China's Yellow River Valley

The agricultural revolution was inspired, in part, by the spreading of domesticated plants and animals and the growth of complex societies. The origin of plant and animal domestication was in China's Yellow River Valley, and the fertile crescent, before it spread in Eurasia. Since Eurasia was connected by land, and there were open trade routes in that region, it was easy for agricultural methods to be adopted by neighbouring communities. The same latitudes of the Eurasian continent meant that plants would grow well in similar climates. This way, they had a productive yield. Either the neighbouring hunter gathers adopted these new methods or they were displaced. The change to the agrarian way of life lead to more developed technology, organized society, and increased populations which required sedentary lifestyles to spread. Therefore the indigenous hunter-gatherers either adapted to this new way of life or else they gradually died off.

Causes of the Neolithic Revolution

Harlan, examining the causes for the Neolithic Revolution, suggests 6 principal reasons which can be summarised to 3 principal categories:

1. Domestication for religious reasons
2. Domestication by crowding and as a consequence of stress
3. Domestication resulting from discovery, based upon the perceptions of food gatherers

With regard to the first explanation, Ian Hodder, who directs the excavations at Çatalhöyük, has suggested that the earliest settled communities, and the Neolithic revolution they represent, actually *preceded* the development of agriculture. He has been developing the ideas first expressed by Jacques Cauvin, the excavator of the Natufian settlement at Mureybet in northern Syria. Hodder believes that the Neolithic revolution was the result of a revolutionary change in the human psychology, a "revolution of symbols" which led to new beliefs about the world and shared community rituals embodied in corpulent female figurines and the methodical assembly of aurochs horns.

An alternative explanation for the origin of agriculture is propounded by Mark Nathan Cohen. Cohen believes that following the widespread extinctions of large mammals in the late Palaeolithic, the human population had expanded to the limits of the available territory and a population explosion led to a food crisis. Agriculture was the only way in which it was possible to support the increasing population on the available area of land.

Consequences of the Neolithic Revolution

Social change

It is often argued that agriculture gave humans more control over their food supply, but this has been disputed by the finding that nutritional standards of Neolithic populations were generally inferior to that of hunter gatherers, and life expectancy may in fact have been shorter. In actual fact, by reducing the necessity for the carrying of children, Neolithic societies had a major impact upon the spacing of children (carrying more than one child at a time is impossible for hunter-gatherers, which leads to children being spaced four or more years apart). This increase in the birth rate was required to offset increases in death rates and required settled occupation of territory and encouraged larger social groups. These sedentary groups were able to reproduce at a faster rate due to the possibilities of sharing the raising of children in such societies. The children accounted for a denser population, and encouraged the introduction of specialization by providing diverse forms of new labour. The development of larger societies called for different means of decision making and led to governmental organization. Food surpluses made possible the development of a social elite who were not otherwise engaged in agriculture, industry or commerce, but dominated their communities by other means and monopolised decision making

Disease

Throughout the development of sedentary societies, disease spread more rapidly than it had during time in which hunter-gatherer societies existed. Inadequate sanitary practices and the domestication of animals may explain the rise in deaths and sickness during the Neolithic Revolution from disease, as diseases jumped from the animal to the human population. Some examples of diseases spread from animals to humans are influenza, smallpox, and measles.

Surprisingly, the humans who first domesticated the wild animals quickly built up immunities to the diseases. Although the humans who built up immunities to the new diseases survived their sickness, others were not so fortunate. According to historians, civilizations which had not domesticated any wild animals nor been exposed to the diseases were not immune at all and "epidemics resulted in which up to 99 percent of the... population was killed".

Subsequent Revolutions

Andrew Sherrat has argued that following upon the Neolithic Revolution was a second phase of discovery that he refers to as the secondary products revolution. Animals, it appears were first domesticated purely as a source of meat. The Secondary Products Revolution occurred when it was recognised that animals also provided a number of other useful products. These included:

- hides and skins (from all domesticated animals)
- manure for soil conditioning (from all domesticated animals)
- wool (from sheep, llama, alpaca and Angora goats)
- milk (from goats, cattle, yaks, sheep and even horses and camels)
- traction (from oxen, buffalo, onagers, donkeys, horses and camels)

Sherrat argues that this phase in agricultural development enabled humans to make use of the energy possibilities of their animals in new ways, and permitted permanent intensive subsistence farming and crop production, and the opening up heavier soils for farming. It also made possible nomadic pastoralism in semi arid areas, along the margins of deserts, and eventually led to the domestication of both the Baktrian and Dromedary camel. Overgrazing of these areas, particularly by herds of goats, greatly extended the areal extent of deserts.

Living in one spot would have more easily permitted the accrual of personal possessions and an attachment to certain areas of land. From such a position, it is argued, prehistoric people were able to stockpile food to survive lean times and trade unwanted surpluses with others. Once trade and a secure food supply were established, populations could grow, and society would have diversified into food producers and artisans, who could afford to develop their trade by virtue of the free time they enjoyed because of a surplus of food. The artisans, in turn, were able to develop technology such as metal weapons. Such relative complexity would have required some form of social organisation to work efficiently and so it is likely that populations which had such organisation, perhaps such as that provided by religion were better prepared and more successful. In addition, the denser populations could form and support legions of professional soldiers. Also, during this time property ownership became increasingly important to all people. Ultimately, Childe argued that this growing social complexity, all rooted in the original decision to settle, led to a second Urban Revolution in which the first cities were built.

The Age of Discovery

In his book *Guns, Germs, and Steel*, Jared Diamond argues that Europeans and East Asians benefited from an advantageous geographical location which afforded them a head start in the Neolithic Revolution. Both shared the temperate climate ideal for the first agricultural settings, both were near a number of easily domesticable plant and animal species, and both were safer from attacks of other people than civilisations in the middle part of the eurasian continent. Being among the first to adopt agriculture and sedentary lifestyles, and neighboring other early agricultural societies with whom they could compete and trade, both Europeans and East Asians were also among the first to benefit from technologies such as firearms and steel swords. In addition, they developed resistances to infectious diseases, such as smallpox, due to their close relationship with domesticated animals. Groups of people who had not lived in proximity with other large mammals, such as the Australian Aborigines, were more vulnerable to infection.

During and after the Age of Discovery, European explorers, such as the Spanish conquistadors, encountered other groups of people which had never or only recently adopted agriculture. Due in part to their head start in the Neolithic Revolution, the Europeans were able to use their technology and endemic diseases, to which indigenous populations had never been exposed, to colonize most of the globe.

Green Revolution

The term Green Revolution is used to describe the transformation of agriculture in many developing nations that led to significant increases in agricultural production between the 1940s and 1960s. This transformation occurred as the result of programs of agricultural research, extension, and infrastructural development largely funded by the Rockefeller Foundation, the Ford Foundation, and national governments. The term "Green Revolution" was first used in 1968 by former USAID director William Gaud, who stated, "[The rapid spread of modern wheat and rice varieties throughout Asia] and other developments in the field of agriculture contain the makings of a new revolution... I call it a Green Revolution based on the application of science and technology.

The Green Revolution has had major social and ecological impacts, which have drawn intense praise and equally intense criticism.

History

Green Revolution in Mexico

The Green Revolution began in 1943 with the establishment of the Office of Special Studies. The venture was a collaboration between the Rockefeller Foundation and the presidential administration of Manuel Avila Camacho in Mexico. While Camacho's predecessor Cárdenas promoted peasant subsistence agriculture through policies of land reform, Avila Camacho's primary goal for Mexican agriculture was to aid in the nation's industrial development and economic growth. US Vice President Henry Wallace, who was instrumental in convincing the Rockefeller Foundation to work with the Mexican government in agricultural development, saw Camacho's ambitions as beneficial to US economic and military interests.

J. George Harrar, who would later become president of the Rockefeller Foundation, headed the Office of Special Studies. Its lead scientists included Norman Borlaug, Edwin Wellhausen, and William Colwell. Researchers from both the United States and Mexico were involved in this program. The main initiative of the Office was the development of high-yielding maize and wheat varieties. Borlaug received the 1970 Nobel Peace Prize for his work on wheat breeding.

The Mexican national government invested heavily in rural infrastructure development, and the adoption of new seed varieties became widespread. Mexico became self-sufficient in wheat production by 1951 and began to export wheat thereafter.

International spread of the Green Revolution

With the experience of agricultural development judged as a success by many of the powerholders involved, the Rockefeller Foundation sought to spread the Green Revolution to other nations. The Office of Special Studies in Mexico became an informal international research institution in 1959, and in 1963 it formally became CIMMYT (The International Maize and Wheat Improvement Center).

The second nation to which the Green Revolution spread was India. The Ford Foundation had a presence in the nation, and their social scientists had decided that the technological development of agriculture was important to the future of India. At the same time Chidambaram Subramaniam, the former Indian Minister of Steel and Mines, became Minister of Food and Agriculture. The Foundation and Indian government collaborated to import a huge amount of wheat seed from CIMMYT. India then began its own Green Revolution program of plant breeding, irrigation development, and financing of agrochemicals. By the late 1970s, the Green Revolution raised rice yields in India by 30 percent and bought India the vital time to curb its population growth without suffering a recurrence of the devastating famines of the 1940s.

The Rockefeller and Ford Foundation jointly established IRRI (The International Rice Research Institute) in the Philippines in 1961. HYVs (high yielding varieties) spread throughout that country, Indonesia, Pakistan, Sri Lanka, and other non-Soviet bloc countries throughout Latin American, Asia, and North Africa. USAID became involved in subsidizing rural infrastructure development and fertilizer shipments.

Agricultural Production and Food Security

Technologies

The projects within the Green Revolution spread technologies that had already existed, but had not been widely used outside of industrialized nations. These technologies included pesticides, irrigation projects, and synthetic nitrogen fertilizer.

The novel technological development of the Green Revolution was the production of what some referred to as "miracle seeds.". Scientists created strains of maize, wheat, and rice that are generally referred to as HYVs or "high yielding varieties." HYVs have an increased nitrogen-absorbing potential compared to other varieties. Since cereals that absorbed extra nitrogen would typically lodge, or fall over before harvest, semi-dwarfing genes were bred into their genomes. Norin 10 wheat, a variety developed by Orville Vogel from Japanese dwarf wheat varieties, was instrumental in developing Green Revolution wheat cultivars. IR8, the first widely implemented HYV rice to be developed by IRRI, was created through a cross between an Indonesian variety named "Peta" and a Chinese variety named "Dee Geo Woo Gen."

With the availability of molecular genetics in Arabidopsis and rice the mutant genes responsible (*reduced height(rht)*, *gibberellin insensitive (gai1)* and *slender rice (slr1)*) have been cloned and identified as cellular signalling components of gibberellic acid, a phytohormone involved in regulating stem growth via its effect on cell division. Stem growth in the mutant background is significantly reduced leading to the dwarf phenotype. Photosynthetic investment in the stem is reduced dramatically as the shorter plants are inherently more stable mechanically. Assimilates become redirected to grain production, amplifying in particular the effect of chemical fertilisers on commercial yield.

HVs significantly outperform traditional varieties in the presence of adequate irrigation, pesticides, and fertilizers. In the absence of these inputs, traditional varieties may outperform HYVs. One criticism of HYVs is that they were developed as F1 hybrids, meaning they need to be purchased by a farmer every season rather than saved from previous seasons, thus increasing a farmer's cost of production.

Production Increases

Cereal production more than doubled in developing nations between the years 1961 – 1985 Yields of rice, maize, and wheat increased steadily during that period. The production increases can be attributed roughly equally to irrigation, fertilizer, and seed development, at least in the case of Asian rice. Some, however, have challenged the purported production increases of Green Revolution agriculture. Miguel A. Altieri, for example, writes that the comparison between traditional systems of agriculture and Green Revolution has been unfair, because Green Revolution agriculture produces monocultures of cereal grains, while traditional agriculture usually incorporates polycultures. Additionally, some traditional systems of agriculture that were displaced by the Green Revolution such as the chinampas in Mexico or raised-field rice farming in Asia are known to be very highly-productive.

Effects on Food Security

The effects of the Green Revolution on global food security are difficult to understand because of the complexities involved in food systems.

The major purported achievement of the Green Revolution has been that the production increases have helped to avoid widespread famine. It is also often claimed that Green Revolution agriculture is responsible for feeding billions of people. These assertions generally assume some variation of the Malthusian principle of population. Malthusianism has been evident throughout the history of the Green Revolution. The team sent to survey Mexican agriculture in 1941 for the Rockefeller Foundation cited the high birth rate and relative inadequacy of its agriculture as a cause for concern. In 1959 the Ford Foundation carried out a study in India that stated the nation's population would outstrip its food supply by 1966, although validity of its methodology was a subject of criticism. At Borlaug's Nobel acceptance speech he stated, "...we are dealing with two opposing forces, the scientific power of food production and the biologic power of human reproduction."

The world population has grown by about four billion since the beginning of the Green Revolution and most believe that, without the Revolution, there would be yet greater famine and malnutrition than the UN presently attributes to the planet Earth (approximately 850 million people suffering from chronic malnutrition in 2005). The average person in the developing world consumes about 25% more calories per day now than before the Green Revolution.

Increasing food production however is not synonymous with increasing food security, and is only part of a larger equation. For example, Amartya Sen's work has found that large historic famines have not been caused by decreases in food supply, but by socioeconomic dynamics and a failure of public action. There are several claims about how the Green Revolution may have decreased food security for some people. One such claim involves the shift of subsistence-oriented cropland to cropland oriented towards production of grain for export and/or animal feed. For example, the Green Revolution replaced much of the land used for pulses that fed Indian peasants for wheat, which did not make up a large portion of the peasant diet. Also, the pesticides involved in rice production eliminated fish and weedy green vegetables from the diets of Asian rice farmers.

Social Changes

Political Impacts

There is a significant amount of evidence suggesting that the Green Revolution had the effect of weakening socialist movements in many nations. In countries like India, Mexico, and the Philippines, technological solutions were sought as an alternative to expanding agrarian reform initiatives, the latter of which were often linked to socialist politics.

Socioeconomic Impacts

The transition from traditional agriculture in which inputs were generated on-farm to Green Revolution agriculture, which required the purchase of inputs, lead to the widespread establishment of rural credit institutions. Smaller farmers often went into debt, which in many cases result in a loss of rights to their farmland. The increased level of mechanization on larger farms made possible by the Green Revolution removed an important source of employment from the rural economy. Because wealthier farmers had better access to credit and land, the Green Revolution increased class disparities. Because some regions were able to adopt Green Revolution agriculture more readily than others (for political or geographical reasons), interregional economic disparities increased as well.

The new economic difficulties of smallholder farmers and landless farm workers led to increased rural-urban migration. The increase in food production led to a decrease in food prices for urban dwellers, and the increase in urban population increased the potential for industrialization. However, industry was unable to absorb all of the displaced agricultural labor and some cities grew at unsustainable rates.

Globalization

In the most basic sense, the Green Revolution was a product of globalization as evidenced in the creation of international agricultural research centers that shared information, and with transnational funding from groups like the Rockefeller Foundation, Ford Foundation, and USAID. Additionally, the inputs required in Green Revolution agriculture created new markets for seed and chemical corporations, many of which were based in the United States. For example, Standard Oil of New Jersey established hundreds of distributors in the Philippines to sell agricultural packages composed of HYV seed, fertilizer, and pesticides.

Ecological Change

Pesticides

Green Revolution agriculture increased the use of pesticides, which were necessary to limit the high levels of pest damage that inevitably occur in monocultures. Organochlorides, a chemical group of pesticides including DDT and dieldrin that spread with the Green Revolution, does not easily break down in the environment and therefore accumulates through the food chain and spread throughout ecosystems. Other problems with pesticides include the poisoning of farm workers, the contamination of water, and the evolution of resistance in pest organism populations.

Water Issues

Irrigation projects have created significant problems of salinization, waterlogging, and lowering of water tables in certain areas.

Biodiversity

The spread of Green Revolution agriculture affected both agricultural biodiversity and wild biodiversity. There is little argument that the Green Revolution acted to reduce agricultural biodiversity, as it relied upon just a few varieties of each crop. This has led to concerns about the susceptibility of a food supply to pathogens that cannot be controlled by agrochemicals, as well as the permanent loss of many valuable genetic traits bred into cereal varieties over thousands of years. To address these concerns, massive seed banks such as CGIAR's International Plant Genetic Resources Institute have been established.

There are varying opinions about the effect of the Green Revolution on wild biodiversity. One hypothesis speculates that by increasing production per unit of land area, agriculture did not need to expand into new, uncultivated areas to feed a growing human population. A counterhypothesis speculates that biodiversity was sacrificed because traditional systems of agriculture that were displaced have often incorporated practices to preserve wild biodiversity, and because the Green Revolution expanded agricultural development into new areas where it was once unprofitable or too arid.

Nevertheless, the world community has clearly acknowledged the negative impacts of agricultural expansion as the 1992 Rio Treaty, signed by 189 nations, has generated numerous national Biodiversity Action Plans which assign significant biodiversity loss to agriculture's expansion into new domains (Whether the habitat used was an arid region or an uninhabited bog, biotic impacts can be extensive.)

Legacy of the Green Revolution

IRRI and CIMMYT became a part of the CGIAR which was established in 1971. Since that time, CGIAR has added many research centers throughout the world.

CGIAR has responded, at least in part, to criticisms of Green Revolution methodologies. This began in the 1980s, and mainly was a result of pressure from donor organizations. Methods like Agroecosystem Analysis and Farming System Research have been adopted to gain a more holistic view of agriculture. Methods like Rapid Rural Appraisal and Participatory Rural Appraisal have been adopted to help scientists understand the problems faced by farmers and even give farmers a role in the development process.

High Yielding Variety

High-yielding varieties or HYVs is the generic name for a group of genetically enhanced cultivars of crops such as rice, maize and wheat that have an increased growth rate, an increased percentage of usable plant parts or an increased resistance against crop diseases. Those crops formed the basis for the Green Revolution. In general, they require a higher level of agricultural care, such as intensive disease control, higher fertilizer levels and controlled water supply. While those crops have enabled the multiplication of agricultural production, their increased demands of fertilizers, pesticides and water control have drawn criticism from environmentalists. Another name is high-response variety, as the high yield requires high input.

High Yielding Varieties vs. Traditional Cultivars

The difference in gross primary production between traditional and high yielding varieties is limited, because until 2005 all attempts to improve the photosynthetic efficiency of any crop have failed. The main explanation of the difference in yield between high yielding varieties and traditional cultivars is the increased rate in which primary plant production is redirected to harvestable products, such as grains. High-yielding varieties of food grains in general have much shorter stems than traditional cultivars, which solves two problems: excessive investment of primary production in the stem and eliminating the balancing problem, which limits the weight of the halm and limits maximum production in traditional cultivars. In addition, HYV cultivars respond favourably to high fertilizer gifts, e.g. nitrogen. This is achieved, because a HYV invests gross photosynthetic production in life cell material rather than in woody structural materials, thus allowing for an increased vegetative growth. As a consequence of their increased growth rate, HYV crops lack an important plant defence mechanism against predatory animals or fungi and as a result are more prone to pests and diseases than traditional cultivars, thus requiring both intensive fertilizer input and high levels of pesticides.

A second characteristic of high-yielding varieties is their homogeneity. In general, HYVs are a cross between two inbred lines which produces a homogenous genotype as offspring, a so-called F1 hybrid, which is genetically homogenous. This makes F1 hybrid HYVs very suitable for mechanical crop care and processing and makes it likier that a certain choice of management parameters will approach the optimum combination for an average plant, thus increasing yield. A drawback of the absence of genetic variation in F1 hybrid HYV crops is the alarming rate in which pests and plant diseases can spread. The high variation in genetic characteristics between plants in a traditional variety make it less likely that a particular pest can attack all plants, but makes it less easy to protect all plants against infestations.

High-yielding varieties often have certain kinds of disease resistance built in, such as resistance against rust, blight or other fungus diseases which reduce the leaf surface and hence hamper the primary production of crops. Because of their homogeneity, disease resistances can be crossed in the population with ease.

Importance of Green Revolution

The world's worst recorded food disaster happened in 1943 in British-ruled India. Known as the Bengal Famine, an estimated four million people died of hunger that year alone in eastern India (that included today's Bangladesh). The initial theory put forward to 'explain' that catastrophe was that there as an acute shortfall in food production in the area. However, Indian economist Amartya Sen (recipient of the Nobel Prize for Economics, 1998) has established that while food shortage was a contributor to the problem, a more potent factor was the result of hysteria related to World War II which made food supply a low priority for the British rulers. The hysteria was further exploited by Indian traders who hoarded food in order to sell at higher prices.

Nevertheless, when the British left India four years later in 1947, India continued to be haunted by memories of the Bengal Famine. It was therefore natural that food security was a paramount item on free India's agenda. This awareness led, on one hand, to the Green Revolution in India and, on the other, legislative measures to ensure that businessmen would never again be able to hoard food for reasons of profit.

However, the term "Green Revolution" is applied to the period from 1967 to 1978. Between 1947 and 1967, efforts at achieving food self-sufficiency were not entirely successful. Efforts until 1967 largely concentrated on expanding the farming areas. But starvation deaths were still being reported in the newspapers. In a perfect case of Malthusian economics, population was growing at a much faster rate than food production. This called for drastic action to increase yield. The action came in the form of the Green Revolution.

The term "Green Revolution" is a general one that is applied to successful agricultural experiments in many

Third World countries. It is NOT specific to India. But it was most successful in India.

What was the Green Revolution in India?

There were three basic elements in the method of the Green Revolution:

1. Continued expansion of farming areas;
2. Double-cropping existing farmland;
3. Using seeds with improved genetics.

Continued Expansion of Farming Areas

As mentioned above, the area of land under cultivation was being increased right from 1947. But this was not enough in meeting with rising demand. Other methods were required. Yet, the expansion of cultivable land also had to continue. So, the Green Revolution continued with this quantitative expansion of farmlands. However, this is NOT the most striking feature of the Revolution.

Double-cropping Existing Farmland

Double-cropping was a primary feature of the Green Revolution. Instead of one crop season per year, the decision was made to have two crop seasons per year. The one-season-per-year practice was based on the fact that there is only natural monsoon per year. This was correct. So, there had to be two "monsoons" per year. One would be the natural monsoon and the other an artificial 'monsoon.'

The artificial monsoon came in the form of huge irrigation facilities. Dams were built to arrest large volumes of natural monsoon water which were earlier being wasted. Simple irrigation techniques were also adopted.

Using Seeds with Superior Genetics

This was the scientific aspect of the Green Revolution. The Indian Council for Agricultural Research (which was established by the British in 1929 but was not known to have done any significant research) was re-organized in 1965 and then again in 1973. It developed new strains of high yield value (HYV) seeds, mainly wheat and rice but also millet and corn. The most noteworthy HYV seed was the K68 variety for wheat. The credit for developing this strain goes to Dr. M.P. Singh who is also regarded as the hero of India's Green revolution.

Economic results of the Green Revolution

1. Crop areas under high-yield varieties needed more water, more fertilizer, more pesticides, fungicides and certain other chemicals. This spurred the growth of the local manufacturing sector. Such industrial growth created new jobs and contributed to the country's GDP.
2. The increase in irrigation created need for new dams to harness monsoon water. The water stored was used to create hydro-electric power. This in turn boosted industrial growth, created jobs and improved the quality of life of the people in villages.
3. India paid back all loans it had taken from the World Bank and its affiliates for the purpose of the Green Revolution. This improved India's creditworthiness in the eyes of the lending agencies.
4. Some developed countries, especially Canada, which were facing a shortage in agricultural labour, were so impressed by the results of India's Green Revolution that they asked the Indian government to supply them with farmers experienced in the methods of the Green Revolution. Many farmers from Punjab and Haryana states in northern India were thus sent to Canada where they settled (That's why Canada today has many Punjabi-speaking citizens of Indian origin). These people remitted part of their incomes to their relatives in India. This not only helped the relatives but also added, albeit modestly, to India's foreign exchange earnings.

Sociological results of the Green Revolution

The Green Revolution created plenty of jobs not only for agricultural workers but also industrial workers by the creation of lateral facilities such as factories and hydro-electric power stations as explained above.

Political results of the Green Revolution

1. India transformed itself from a starving nation to an exporter of food. This earned admiration for India in the comity of nations, especially in the Third World.
2. The Green Revolution was one factor that made Mrs. Indira Gandhi (1917-84) and her party, the Indian National Congress, a very powerful political force in India (*it would however be wrong to say that it was the only reason*).

Statistical Results of the Green Revolution

1. The Green Revolution resulted in a record grain output of 131 million tons in 1978-79. This established India as one of the world's biggest agricultural producers. No other country in the world which attempted the Green

Revolution recorded such level of success. India also became an exporter of food grains around that time.

2. Yield per unit of farmland improved by more than 30 per cent between 1947 (when India gained political independence) and 1979 when the Green Revolution was considered to have delivered its goods.
3. The crop area under HYV varieties grew from seven per cent to 22 per cent of the total cultivated area during the 10 years of the Green Revolution. More than 70 per cent of the wheat crop area, 35 per cent of the rice crop area and 20 per cent of the millet and corn crop area, used the HYV seeds.

Limitations of the Green Revolution

1/ Even today, India's agricultural output sometimes falls short of demand. The Green Revolution, howsoever impressive, has thus NOT succeeded in making India totally and permanently self-sufficient in food. In 1979 and 1987, India faced severe drought conditions due to poor monsoon; this raised questions about the whether the Green Revolution was really a long-term achievement. In 1998, India had to import onions. Last year, India imported sugar.

However, in today's globalised economic scenario, 100 per cent self-sufficiency is not considered as vital a target as it was when the world political climate was more dangerous due to the Cold War.

2. India has failed to extend the concept of high-yield value seeds to all crops or all regions. In terms of crops, it remain largely confined to foodgrains only, not to all kinds of agricultural produce. In regional terms, only Punjab and Haryana states showed the best results of the Green Revolution. The eastern plains of the River Ganges in West Bengal state also showed reasonably good results. But results were less impressive in other parts of India.
3. Nothing like the Bengal Famine can happen in India again. But it is disturbing to note that even today, there are places like Kalahandi (in India's eastern state of Orissa) where famine-like conditions have been existing for many years and where some starvation deaths have also been reported. Of course, this is due to reasons other than availability of food in India, but the very fact that some people are still starving in India (whatever the reason may be), brings into question whether the Green Revolution has failed in its overall social objectives though it has been a resounding success in terms of agricultural production.
4. The Green Revolution cannot therefore be considered to be a 100 percent success.

British Agricultural Revolution

The British Agricultural Revolution describes a period of agricultural development in Britain between the 16th century and the mid-19th century, which saw a massive increase in agricultural productivity and net output. This in turn supported unprecedented population growth, freeing up a significant percentage of the workforce, and thereby helped drive the Industrial Revolution. How this came about is not entirely clear. In recent decades, historians cited four key changes in agricultural practices, enclosure, mechanization, four-field crop rotation, and selective breeding, and gave credit to a relatively few individuals. However, current thinking suggests that fundamental changes in workforce organization, crops and farming methods were largely responsible. For this reason, historians may argue that there was in fact no "revolution", simply a rapid evolution of methods that led to a sharp rise in productivity.

Enclosure

Prior to the 18th century, agriculture had been much the same across Europe since the Middle Ages. The open field system was essentially post-feudal, with each farmer subsistence-cropping strips of land in one of three large fields held in common and splitting up the products likewise.

Beginning as early as the 12th century, some of the common fields in Britain were enclosed into individually owned fields, and the process rapidly accelerated in the 15th and 16th centuries as sheep farming grew more profitable. This led to farmers losing their land and their grazing rights, and left many unemployed. In the 16th and 17th centuries, the practice of enclosure was denounced by the Church, and legislation was drawn up against it; but the developments in agricultural mechanization during the 18th century required large, enclosed fields in order to be workable. This led to a series of government acts, culminating finally in the General Inclosure Act of 1801.

While farmers received compensation for their strips, it was minimal, and the loss of rights for the rural population led to an increased dependency on the Poor law. Surveying and legal costs weighed heavily on poor farmers, who sometimes even had to sell their share of the land to pay for its being split up. Only a few found work in the (increasingly mechanised) enclosed farms. Most were forced to relocate to the cities to try to find work in the emerging factories of the Industrial Revolution. By the end of the 19th century the process of enclosure was largely complete.

Mechanization

Jethro Tull made the first advancements in agricultural technology with his seed drill (1701)—a mechanical seeder which distributed seeds efficiently across a plot of land. However, he was not the first to invent a seed drill. It took a century after the publication in 1731 of his *Horse hoeing husbandry* for farmers to widely adopt the technology.

Joseph Foljambe's Rotherham plough (1730), while not the first iron plough, was the first iron plough to have any commercial success, combining a number of technological innovations in its design, and being lighter than traditional ploughs. It remained in use in Britain until the development of the tractor.

Andrew Meikle's threshing machine of 1786 was the final straw for many farm labourers, and led to the 1830 agricultural rebellion of Captain Swing (a mythical character comparable to the Luddite's Ned Ludd). Increasing mechanisation improved farming efficiency and reduced costs.

Four-field Crop Rotation

During the Middle Ages, the open field system had employed a three year crop rotation, with a different crop in each of the three fields, eg. wheat and barley in two, with the third fallow. 'Fallow' is a term which means that the field is empty, there is nothing growing there. Over the following two centuries, the regular planting of nitrogen-rich legumes in the fields which were previously to fallow slowly increased the fertility of croplands. The planting of legumes (leguminosae, plants of the pea/ bean family) helped to increase plant growth in the empty field because they used a different set of nutrients to grow than the grains. The legumes put back nutrients the grains used, nitrates produced from nitrogen in the atmosphere, and the grains put back the minerals the legumes used. In a way, they fed each other. When the pastures were brought back into crop production after their long fallow, their fertility was much greater than they had been in medieval times. The farmers in Flanders (current day Belgium), however, discovered a still more effective four-field rotation system, introducing turnips and clover to replace the fallow year. Clover was both an ideal fodder crop, and it actually improved grain yields in the following year (clover is part of the pea family, leguminosae). The improved grain production simultaneously increased livestock production. Farmers could grow more livestock because there was more food, and manure was an excellent fertili

Scottish Agricultural Revolution

The Agricultural Revolution in Scotland began in the mid-eighteenth century with the improvements of the lowland farmland and the beginning of a transformation of Scottish agriculture from one of the most backward into what was to become the most modern and productive system in Europe.

The traditional system of agriculture in lowland Scotland had existed unchanged for hundreds of years. In many ways it was a totally rural economy, the land being worked by the cottars on the centuries-old *runrig* system of subsistence farming.

The Agricultural Revolution led directly to what is increasingly becoming known as the Lowland Clearances, when hundreds of thousands of cottars and tenant farmers from central and southern Scotland were, in many cases, forcibly moved from the farms and small holdings they had occupied for hundreds of years.

Many small settlements were torn down, their occupants forced either to the new purpose-built villages built by the landowners such as John Cockburn of Ormiston to house the displaced cottars on the outskirts of the new ranch-style farms, or to the new industrial centres of Glasgow, Edinburgh, or northern England.

There, some found employment in the new industries of cotton or linen weaving. Tens of thousands of others emigrated to the newly opened-up colonies of Canada and the northern states of the United States, finding opportunities there to own and farm their own land.

Many of the recently displaced Scots found a new home in Nova Scotia (latin for New Scotland) and would remain there for generations. Even today, the majority of Nova Scotians have ancestors that originated in Scotland.

Some chose to remain on the land, either by choice or out of sheer necessity, but rents were increased to the extent that tenant or sub-tenant were eventually forced to sell. Consequently, the cottars and their way of life disappeared altogether in many parts of lowland Scotland]].

However, in time, the improvements in agriculture in the lowlands, the so-called Age of Improvement, led to a more stable and modern system of agricultural production which was to make Scotland the envy of Europe.

Horticulture

The Latin words *hortus* (garden plant) and *cultura* (culture) together form horticulture, classically defined as the culture or growing of garden plants. Horticulturists work in plant propagation, crop production, plant breeding and genetic engineering, plant biochemistry, plant physiology, and the storage, processing, and transportation of fruits, berries, nuts, vegetables, flowers, trees, shrubs, and turf. They improve crop yield, quality, nutritional value, and resistance to insects, diseases, and environmental stresses. Genetics is also used as a valuable tool in the development of plants that can synthesize chemicals for fighting disease (including cancers).

Horticulture involves five areas of study. These areas are floriculture (includes production and marketing of floral crops), landscape horticulture (includes production, marketing and maintenance of landscape plants), olericulture (includes production and marketing of vegetables), pomology (includes production and marketing of fruits), and postharvest physiology (involves maintaining quality and preventing spoilage of horticultural crops).

Horticulturists can work in industry, government, or educational institutions. They can be cropping systems engineers, wholesale or retail business managers, propagators and tissue culture specialists (fruits, vegetables, ornamentals, and turf), crop inspectors, crop production advisors, extension specialists, plant breeders, research scientists, and of course, teachers.

College courses that complement Horticulture are biology, botany, entomology, chemistry, mathematics, genetics, physiology, statistics, computer science, and communications, garden design, planting design. Plant science and horticulture courses include: plant materials, plant propagation, tissue culture, crop production, post-harvest handling, plant breeding, pollination management, crop nutrition, entomology, plant pathology, economics, and business. Some careers in horticultural science require a masters (MS) or doctoral (PhD) degree.

Horticulture takes place in many garden and plant growth centres. Plants are often grown as seedlings within plant nurseries. Activities in nurseries range from preparing seeds and cuttings to growing fully mature plants. These are often sold or transferred to ornamental gardens or market gardens.

Corporate Farming

Corporate farming is a term that describes the business of agriculture, specifically, what is seen by some as the practices of would-be megacorporations involved in food production on a very large scale. It is a modern food industry issue, and encompasses not only the farm itself, but also the entire chain of agriculture-related business, including seed supply, agrichemicals, food processing, machinery, storage, transport, distribution, marketing, advertising, and retail sales. The term also includes the influence of these companies on education, research and public policy, through their educational funding and government lobbying efforts. *Corporate farming* is often used synonymously with *agribusiness* (although *agribusiness* quite often is not used in the corporate farming sense), and it is seen as the destroyer of the family farm. Two percent of all farms in the United States are owned by corporations or other non-family entities, but only half of those farms earn more than $50,000 per year.

Critics argue that the ultimate goal of corporate farming is to vertically integrate the entire process of food production, from the development of proprietary strains of DNA through to the distribution and sale of food to consumers. Some corporations are considered to be well on the way to achieving this objective, and have become very large in the process, such as Archer Daniels Midland and the privately held Cargill, with 2004 revenues of $62.9 billion.

Corporate farming is a fairly broad term that deals with the general practices and effects of a small number of large, global corporations that dominate the food industry. It does not refer simply to any incorporated agribusiness enterprise, although most agricultural businesses today are in some way economically connected to the dominant food industry players. As such, it may be thought of as a movement, which is at times also referred to as anti-corporate farming.

Effects Ascribed to Corporate Farming

Agriculture is an industry which provides significant economies of scale to large producers. As is the trend in such industries, agricultural production in most free-market economies has been increasingly concentrated in the hands of a small number of large companies. As this trend continues, so too does the risk that markets for food will display monopolistic or oligarchical characteristics. At the same time, competition law and policy seeks to redress such market distortions.

Concentration of Production

Corporate farming is criticized for its tendency to concentrate food production, through the adoption of methods which are designed to maximise crop yield. At the same time, through a process of vertical integration, corporate farming results in the concentration of not only ownership of the means of production, but also the distribution and sale of food which is produced. Over time this has resulted in the steady decrease of the number of farms and the percentage of independent farmers. Those farmers who cannot compete or whose businesses are purchased are no longer able to make independent production decisions, and are forced to sign production contracts with corporations. This is essentially an anti-monopolist criticism.

The trend towards concentration began in the chicken and vegetable industries and has since expanded to hog and grain production. In 1997, some 60% of hogs sold within the U.S. were sold under some form of contract, whereas in 1980 only 5% of hogs were sold in this manner.

As production continues to concentrate and is coupled with increasing reliance on technology, farmers complain about their increasing remoteness from centers of population or production. For example, farm machinery repair services, which were once as close as two miles away, are increasingly as far as 40 miles away.

Food Quality and Cheap Food

A critical consideration is whether the quality of the food which reaches the consumer is as good as it would be under alternative structures of ownership and production. To the extent that corporate farming primarily seeks to maximise yield and profit, this is seen as adversely impacting upon nutritional value, freshness and flavour, as well as upon the range of products available for consumers. Corporate farming practices may also more readily involve the use of genetically modified crops, hormones, preservatives, color additives and insecticides, again with a view to maximising yield and profit.

In the United Kingdom there has been a growing reaction against factory farmed produce in recent years, with people feeling that they can obtain higher quality products (admittedly at a higher price) if they know the provenance (the local and often small scale source) of the food they buy. There is little or no comparative information about flavor and nutritional value compared internationally and over time (i.e. today compared with the past) but there is anecdotal and some scientific evidence that US factory farming may have resulted in a deterioration (and not simply a change) of food quality.

Genetic Engineering

The debate about large corporations in agriculture is heightened at the intersection with the already controversial issue of genetically modified food. Critics of both are concerned that these large businesses will leverage their economic and political power to defeat attempts to regulate or restrain the spread of genetic engineering in agriculture, or leverage intellectual property rights in GMOs to unfair advantage over their competitors, against the interests of users and consumers of their products, and to the detriment of the environment.

A reason given in support of patents is that they provide an economic incentive for the creation of new and useful inventions by temporarily rewarding the inventor with monopoly profits. The patent holder has certain exclusive rights to the exploitation of her invention, which may result in significant economic rewards. However, consumers may choose to buy rival products (which could be protected by other patents), and the lifetime of the patent is limited.

Simple possession of a naturally occurring seed (free from patent restrictions) gives one the ability and the legal right to grow crops from the seed, to modify the breed, and to sell, exchange, or share the seed as one sees fit. With patent rights, however, the inventor (often a large corporation) may choose to restrict how farmers may use a given organism, in order extract the economic value of each type of use, and to extract economic value from each user.

Critics of GMOs and corporations say that these rights give already powerful corporations an even greater advantage, and in a way which disrupts millennia-old agricultural practices. Supporters of genetic engineering emphasize the potential benefits in nutrition, reduced environmental impact, and increased productivity that may be possible with the technology. They say that the additional economic rewards are necessary to encourage the large capital investments needed to make useful advancements in the field.

Corporate Farm vs Family Farm

Farms are expensive to operate; input costs include farm machinery, crop insurance, fertilizers, irrigation, pesticides, fuel, and seeds. Some people question whether small family farms are still economically sustainable in the United States. However, there is a growing resurgence of interest in organic, free range, and locally grown family farm products.

One major difference between independent farming and corporate farming is that a corporate farmer is usually a contracted employee, rather than the owner of the farm. However, ownership itself does not mean independence. An owner-operated farm today faces many constraints that are completely out of the owner's control. Most of these can be seen in light of increasing concentration of ownership, not only of farms, but of the equipment and inputs necessary to farm, and the available sales channels.

Production contracts are a primary means of control and vertical integration of family farms. These are of two general types. Production management contracts specify the methods farmers must use. Resource-providing contracts require the contractor to also provide materials (e.g) and equipment. Under the latter, increasingly prevalent arrangement, the family farm owns its land and "sells" its output, but retains no real decision making control over the essential farming activities, like crop selection, equipment purchase, production methods, sales channels, and buyers.

A prime example is the drive to constantly improve production efficiency, as measured by farm output. By using successive waves of new technology (in agrichemicals, mechanization, crop varieties, drugs, etc), output has steadily risen over the past decades. This in turn has contributed to steadily driving down the price farmers can get for their output, and driving up the cost (to the farmer) of production. As the cost of remaining in production rises, and income falls, only the larger business entities, with the ability to profit from outside of the immediate farming activities (such as through financial services, agrichemical production, food distribution, and so forth) can afford to remain in the game.

Benefits of Corporate Farming

The core argument for the methods criticized as corporate farming is essentially: "This is the way to keep up with population growth, and to make inroads into feeding billions of people to developed nation standards—*this is the only way to feed the world.*" Indeed, rapid technological development and large-scale global production management are responsible for an unprecedented abundance of inexpensive, widely available, attractive, "safe" food. By lowering the cost of raw food inputs, creating sophisticated long-distance distribution networks, producing processed convenience foods, and making food available year-round in vastly stocked supermarkets, corporate farming has presented consumers in the wealthiest regions of the world with an immense variety of food, at relatively low cost. Today, in North America, only about 15% of average income is spent on food. By this measure, providing these methods are sustainable, corporate farming would appear to be a tremendous success.

Crop Rotation

Crop rotation is the practice of growing a series of dissimilar types of crops in the same space in sequential seasons to avoid the buildup of pathogens and pests that often occurs when one species is continuously cropped. Crop rotation also seeks to balance the fertility demands of various crops to avoid excessive depletion of soil nutrients. A traditional component of crop rotation is the replenishment of nitrogen through the use of green manure in sequence with cereals and other crops. It is one component of polyculture. Crop rotation can also improve soil structure and fertility by alternating deep-rooted and shallow-rooted plants.

History

Crop rotation was already mentioned in the Roman literature, and referred to by great civilizations in Africa and Asia. From the end of the Middle Ages until the 20th century, the three-year rotation was practiced by farmers in Europe with a rotation of rye or winter wheat, followed by spring oats or barley, then letting the soil rest (fallow) during the third stage. The fact that suitable rotations made it possible to restore or to maintain a productive soil has long been recognized by planting spring crops for livestock in place of grains for human consumption. Crop rotation in the fourteenth century was considerably more widespread after John Lloyd invented the patent crop rotator. A four-field rotation was pioneered by farmers, namely in the region Waasland in the early 16th century and popularised by the British agriculturist Charles Townshend in the 18th century. The system (wheat, barley, turnips and clover), opened up a fodder crop and grazing crop allowing livestock to be bred year-round. The four-field crop rotation was a key development in the British Agricultural Revolution.

Contrary to the widespread myth, crop rotation was not pioneered in the United States by George Washington Carver. Carver merely taught the standard technique to poor black farmers as part of his extension education program.

In the Green revolution, the practice of crop rotation gave way in some parts of the world to the practice of simply adding the necessary chemical inputs to the depleted soil, e.g., replacing organic nitrogen with ammonium nitrate or urea and restoring soil pH with lime. However, disadvantages of monoculture from the standpoint of sustainable agriculture have since become apparent.

Method and Purpose

Crop rotation avoids a decrease in soil fertility, as growing the same crop repeatedly in the same place eventually depletes the soil of various nutrients. A crop that leaches the soil of one kind of nutrient is followed during the next growing season by a dissimilar crop that returns that nutrient to the soil or draws a different ratio of nutrients, for example, rices followed by cottons. By crop rotation farmers can keep their fields under continuous production, without the need to let them lie fallow, and reducing the need for artificial fertilizers, both of which can be expensive.

Legumes, plants of the family Fabaceae, for instance, have nodules on their roots which contain nitrogen-fixing bacteria. It therefore makes good sense agriculturally to alternate them with cereals (family Poaceae) and other plants that require nitrates. A common modern crop rotation is alternating soybeans and maize (corn). In subsistence farming, it also makes good nutritional sense to grow beans and grain at the same time in different fields.

Crop rotation is also used to control pests and diseases that can become established in the soil over time. Plants within the same taxonomic family tend to have similar pests and pathogens. By regularly changing the planting location, the pest cycles can be broken or limited. For example, root-knot nematode is a serious problem for some plants in warm climates and sandy soils, where it slowly builds up to high levels in the soil, and can severely damage plant productivity by cutting off circulation from the plant roots. Growing a crop that is not a host for root-knot nematode for one season greatly reduces the level of the nematode in the soil, thus making it possible to grow a susceptible crop the following season without needing soil fumigation.

It is also difficult to control weeds similar to the crop which may contaminate the final produce. For instance, ergot in weed grasses is difficult to separate from harvested grain. A different crop allows the weeds to be eliminated, breaking the ergot cycle.

This principle is of particular use in organic farming, where pest control may be achieved without synthetic pesticides.

A general effect of crop rotation is that there is a geographic mixing of crops, which can slow the spread of pests and diseases during the growing season. The different crops can also reduce the effects of adverse weather for the individual farmer and, by requiring planting and harvest at different times, allow more land to be farmed with the same amount of machinery and labor.

The choice and sequence of rotation crops depends on the nature of the soil, the climate, and precipitation which together determine the type of plants that may be cultivated. Other important aspects of farming such as crop marketing and economic variables must also be considered when choosing a crop rotation.

Short Rotation Coppice

Species Used

SRC uses high yield varieties of poplar and willow. Typically the willow species chosen are varieties of the Common Osier or Basket Willow, *Salix viminalis*. Poplar is generally planted for visual variation rather than being a commercial crop, although some varieties can outperform willow on suitable sites. Species are selected for their acceptance of varying climate and soil conditions, relative insusceptibility to pests and diseases, ease of propagation and speed of vegetative growth. To combat pests such as brassy and blue willow beetles, as well as the fungal pathogen *Melampsora* (a rust), planting a carefully selected mix of varieties is recommended.

Planting

SRC can be planted on a wide range of soil types from heavy clay to sand including land reclaimed from gravel extraction and colliery spoil. Where used as a pioneer species the SRC yield may be smaller. Water availability to the roots is a key determinant for the success of the SRC. Saplings are planted at a high density, around 15,000 per hectare for willow and 12,000 per hectare for poplar. Planting takes place around March to take advantage of the high moisture of the soil in the spring and the amount of sunshine in the early summer. The most efficient planting machines plant four rows at a time and can plant a hectare in around three hours. Saplings are left to grow for a year and then coppiced. The first three years are part of the establishment phase and do not yield much dry matter.

Harvesting

After four years the plantation will be ready for harvest. Harvests take place on a two to five year cycle, shorter for willow and longer for poplar, and are carried out in winter after leaf fall when the soil is frozen. The established root system and the nutrients stored in the roots and stumps guarantee vigorous growth for the shoots. A plantation will yield from 8 to 18 tonnes of dry woodchip per hectare per year. A plantation can be harvested for up to thirty years before needing to be replanted.

When willow or poplar shoots are harvested as whole stems they are easy to store. The stems can be dried for combustion in a pile outdoors; the moisture content of the wood will decrease to about 30 % on average until the next autumn. The stems can be cut further into billets that may not need to be chipped depending on use.

Where wood chip is being produced it is most efficient to use direct-chip harvesters. These are heavy self-powered machines that cut and chip the shoots on a loading platform. Some can be attached to a normal tractor and a hectare can be harvested in around 3 hours. Direct chipping reduces costs as a separate chipping in the store will not be needed; however, the wood chip needs to be well stored to avoid it composting. Harvesting Poplar requires heavier machinery as it produces fewer and heavier stems.

The price of dry willow as a heating fuel is currently around 45 euro per tonne in most of Europe. This is not a relatively high-return crop, but it is low-maintenance and is a way of utilising difficult fields. Small scale production can be combined with the production of material for wicker work. Correctly managed there is little need for pesticides or treatments.

Greenhouse Gas Impact

SRC has a low greenhouse gas impact as any carbon dioxide released in power generation will have been sequestered by the plantation over just a few years. Some carbon may also be stored in the soil, however this is dependent on the carbon content of the soil to begin with.

The carbon costs associated with SRC are: the planting, farming and chipping of the SRC plantation, generally done with fossil fuel powered machinery; the crops require herbicides during establishment, fertiliser throughout growth, and occasional pesticide treatment — these chemicals require substantial amounts of energy and potential fossil fuel usage through manufacture.

Electricity or heat from SRC provides between three and six times the CO_2 reduction per pound that can be obtained from bioethanol from cereal crops. However, the reduction in CO_2 emissions is slightly lower than grass energy crops such as *Miscanthus* grass due to higher maintenance costs.

Environmental Impacts

Good conservation management encouraging biodiversity can reduce the reliance on pesticides. Biomass crops such as SRC willow show higher levels of biodiversity in comparison with intensive arable and grassland crops. SRC has a higher water consumption than agricultural crops. The root systems of SRC have a lower impact on archaeological remains than forestry but greater than agricultural crops such as wheat.

Energy Generation

A power station requires around 100 hectares (1 km^2) of SRC for 1 MW of power capacity. The primary barrier to establishing plantations is the cost as there is no financial reward for four years from a large initial investment. The current nature of the power industry generally requires flexibility in energy supply which is incompatible with the long term commitment SRC requires; however, there is much interest in SRC due to the need to reduce fossil carbon emissions. Grants may also be available in some jurisdictions to further this type of land-use.

Alternative Crops in Double-Crop Systems for Missouri: A Case Study

Double cropping increases the amount of time land is used for crop production and can increase potential profit. There are ecological as well as economic advantages to increasing the amount of time the land is used for production. A winter grain crop, traditionally wheat, can act as a cover crop to prevent erosion. A warm-season double crop, often soybean, may yield smaller returns than a full-season crop but often averages a positive net return. This high-intensity cropping system may reduce disease and pest incidence by breaking pest cycles.

There is a need to identify alternative crops that can fit easily into double-crop systems with little or no added equipment costs and with minimal change in management. Acceptable alternative crops will benefit farmers through decreased incidence of pests and diseases, improved soil erosion control, and reduced economic risk for the producer.

In a two-year study at the University of Missouri, five alternative crops were evaluated agronomically and economically based on their potential to perform well in a double-crop system in central and southern Missouri. Double crops planted after the harvest of canola or wheat or after a fallow period were evaluated at three locations throughout the state between 1992 and 1994. The agronomic and economic performance of soybean was compared with that of four alternative double crops: amaranth, buckwheat, pearl millet and sunflower.

Uses and Markets for Alternative Crops

Canola

Canola is a type of rapeseed. It differs from rapeseed in that it has lower levels of erucic acid and glucosinolates, both of which are undesirable in foods. Canola is also low in saturated fats, which makes it a good source of vegetable oil for human consumption. Rising consumer demand for canola oil is not being met by domestic canola production, often because of added costs for transportation between widely separated collection points.

Wheat typically yields more than canola, but has a lower value per unit of grain. The canola crop is similar to wheat in its growth and management, although canola is occasionally subject to winter kill in Missouri. Prices for canola fluctuate depending on prices of other vegetable oils, especially soybean.

Amaranth

Amaranth is referred to as a pseudocereal because it has some of the same uses as true cereal grains but differs in plant and seed type. It is high in the amino acids lysine and tryptophan, which are typically low in cereal grains. Amaranth is currently used in the specialty food market as an added ingredient in breakfast cereals, breads and crackers and has potential as an additive in animal feeds or for industrial use. Most amaranth undergoes processing into flour, but some grain is marketed directly by farmers to consumers.

Current markets for amaranth are small, and only a few companies process the grain. Production practices and costs for amaranth are similar to those for sunflower.

Prices for amaranth are higher than those for other grains, provided a producer has a buyer or a marketing arrangement. However, undesirable agronomic characteristics of amaranth, such as shattering and lodging, indicate the need for further breeding efforts to improve the crop.

Buckwheat

Buckwheat is a pseudocereal that matures nine to ten weeks after emergence, making it an excellent option for double cropping in Missouri. Buckwheat is commonly known as a flour component in pancakes and baked goods. Buckwheat is also a primary ingredient in soba-style noodles made in Japan.

Production practices for buckwheat are similar to those for double-crop soybean. The crop is typically planted in narrow rows and is combine harvested with a grain head. Prices for buckwheat are slightly higher than for soybean on a per unit weight basis, but yields may be lower. Producers in Missouri are obtaining a positive net return despite added transportation costs. Buckwheat is increasingly recommended as a double-crop for Missouri in response to solid and expanding domestic and export markets.

Pearl Millet

Pearl millet was domesticated in Africa for human consumption and has slightly better food value than either wheat or rice. Pearl millet is a drought-tolerant crop that shows a reduction in yield with a shortened growing season. Production practices for pearl millet are similar to those for soybean, except that pearl millet requires an application of nitrogen and may require cultivation for weed control because of a lack of labeled herbicides.

Pearl millet has been used in North America primarily as a forage crop but is finding increasing use as a grain crop. In Georgia, pearl millet is being used by large-scale poultry operations as a component of poultry rations. Unfortunately, the price of pearl millet is low and does not reflect the grain's value relative to its nutritional quality.

Sunflower

Sunflower is typically grown in Missouri as a component in birdseed mixes. Sunflower can be planted and managed with methods and equipment similar to those used for corn. Future markets for sunflower include expanded processing of specialty varieties for use as a component in snack foods, birdseed mixes, baked goods and specialty oils. In addition, oil extraction and confectionery byproducts can be fed to livestock. Sunflower has slightly higher production costs than soybean, primarily because of the added nitrogen and equipment necessary with sunflower. Price of sunflower fluctuates with the price of soybean and other vegetable oils.

Premiums are paid for specialty sunflower, such as confectionery seeds, or varieties with oil high in oleic acid. In-state bird feed producers provide an important market for sunflower in Missouri.

Winter-crop Management

Winter wheat and canola require similar methods of management. Winter canola, like winter wheat, is seeded in the fall and goes dormant over winter. Growth of canola resumes from living crown tissue in the spring.

Varieties

The best varieties of winter wheat and winter canola for double-crop production systems are those that consistently produce high yields and yet mature early enough to allow for timely planting of the following double crop. An important feature of a successful conola variety is resistance to winter kill.

Planting and Fertilization

Seedbeds for canola and wheat should be prepared using methods typical for small-seeded grain production. Sites are either plowed or chiseled and then cultivated twice. Soil samples should be taken before planting winter crops at all sites and analyzed for pH and nutrient content. Nitrogen, phosphorus and potassium are applied before planting. Spring topdressed applications of nitrogen are made in March. Canola is typically planted in early to mid-September, a month earlier than wheat.

Harvest

Winter crops are harvested with a combine in mid-June to early July. Canola harvest needs to be timely because of the potential for reduced yields due to seed shatter loss if pods become too dry.

Economic Considerations

Although the alternative crops discussed here fit into Missouri's double-crop season, to be acceptable, they must also be economically competitive with the traditional double crop. The alternative crops in the Missouri study tend to show greater variations in grain yield than soybean, which is well adapted to the area and for which appropriate management has been optimized. To reflect possible yield variation, the Missouri study considered three alternative yield levels for evaluating the economic suitability of these double-crop options: average yield from field tests, base yield, and optimum yield. Average grain yields from field tests for all crops except wheat were based on yields from the three sites in the Missouri study. "Base" yields of winter and double crops reflect the minimum estimated yield a producer in Missouri could expect in most years, assuming a good crop stand and no catastrophic events. "Optimum" yields for winter and double crops reflect the highest replicated yield values from tests conducted in this study.

TABLE 2

Grain yields in pounds per acre of five double crops, planted after wheat or canola.

Average crop yield	*Double crop*	*Winter crop (pounds per acre)*
Amaranth	canola	891
	wheat	851
Buckwheat	canola	972
	wheat	892
Pearl millet	canola	1,015
	wheat	458
Soybean	canola	1,345
	wheat	1,455
Sunflower	canola	1,832
	wheat	1,623

Variable and fixed costs were considered in determining costs of production for ten double-crop systems. Since current regulations limit pest control options for alternative crops, estimated weed control costs among the double crops varied. No herbicides are labeled for use in amaranth and buckwheat. Some herbicides are labeled for use on soybean, sunflower and pearl millet; costs for one application of a preemergent or postemergent labeled herbicide were assigned to these crops. Herbicide application costs were taken from experience in the two-year study. Because several insect pests are known to cause significant damage to double-crop sunflower in Missouri, one application of insecticide was assigned to this crop.

Since wide rows allow for cultivation in amaranth, this crop was assessed an additional $3.25 per acre in labor and operating machinery costs to reflect the use of increased cultural controls. Mechanical weed control options in buckwheat are limited because the crop is planted in narrow rows, so no additional labor charge was assessed to buckwheat.

Double-crop Management

No-till methods can be used successfully when planting alternative double crops. However, canola and wheat residues can occasionally clog furrows during planting, making them difficult to close. Canola stubble is lighter and has less volume than wheat residues and generally allows easier no-till planting. Some producers have reported better yields of soybean when planted after canola than when planted after wheat. University of Missouri test results indicate, however, that the effects of canola and wheat residues on following double crops are neither consistently positive nor consistently negative. In general, wheat residues, being thicker and slower to break down than canola residue, could be an advantage in reducing erosion and preserving topsoil moisture. However, wheat residues often complicate no-till planting and seedbed preparation by clogging equipment and furrows. Both volunteer wheat and volunteer canola can cause serious yield losses in sequential crops and must be managed with cultural or chemical controls. Varieties. Improved commercial varieties are available for all of the double crops evaluated in this study. Varieties should be carefully selected to achieve the highest return in local markets. Resistence to pests is also a significant consideration for alternative double crops, as few pesticides are labeled for use with these crops. Planting and fertilization. Because double crops cannot be planted until after wheat or canola harvest in late June or early July, yields for most of these double crops are lower than they would be if the crop were planted for optimum growth. A later planting date typically results in lowered grain yield because of inadequate summer rainfall and a shortened growth and maturation period. The optimum planting time for buckwheat, unlike that for other crops in this study, occurs from mid-July to early August. The nine- to ten-week growing period for buckwheat gives the producer more flexibility, because buckwheat can be planted weeks after the harvest of wheat or canola and still produce maximum grain yield. Nutrient application rates for double crops should be adjusted downward from those for full-season crops to reflect the lower yield potential (the exception being buckwheat). Producers should follow recommendations based on soil test results and realistic yield expectations.

Canola and Wheat

Results of the study show that canola can be profitable in double-crop systems in central and southern Missouri at yields above base levels, although less so than wheat. Analysis of canola in this study does not incorporate the risk associated with the higher rate of winter kill, but this must be a factor in a producer's decision to grow canola. Improved management techniques and varieties are gradually reducing the risk of canola winter kill.

Amaranth

The high grain price for amaranth resulted in amaranth double-crop systems having the highest potential net return of the double-crop options. Moreover, double-crop amaranth was highly competitive with double-crop soybean on the basis of projected net return. Considering all yield estimates, double-croppped amaranth had net returns 1.5 to 10 times that of double-crop soybean after canola or wheat. Greater labor and transportation costs were included in the variable costs for double-crop amaranth production. An estimated 20 hours of labor per acre is required for amaranth cultivation and harvest. Increases in amaranth production may lower prices to $1.10 per pound, which would reduce the net returns estimated here by up to 50 percent. However, returns would still be above the net return for double-crop soybean at average and optimum yield levels.

Lack of market demand for amaranth in the United States has limited production to less than 5,000 acres annually. Consequently, amaranth is an option for only a small number of producers at this time.

Buckwheat

Buckwheat double-crop systems had net returns at average and optimum yield potentials that were lower than those for amaranth, soybean and sunflower. However, at base yield levels, net return for the double-crop systems including buckwheat were slightly greater than those involving soybean. In a comparison of production costs, the main differences with soybeans are fertilizer and herbicide costs. Buckwheat receives an application of nitrogen that soybean does not, but a herbicide application on soybean makes total costs of production for the two double-crops similar.

TABLE 3

Net profit or loss for ten double-crop options at study yield levels, in dollars per acre.

Double crop system	*Net return*		
	Base yield	*Average yield*	*Optimum yield*
Wheat-amaranth	117.11	404.33	740.82
Wheat-buckwheat	12.08	120.82	183.51
Wheat-soybean	10.48	171.53	323.11
Wheat-sunflower	14.74	185.51	267.19
Wheat-pearl millet	17.84	57.97	143.38
Canola-amaranth	76.89	334.59	662.48
Canola-buckwheat	-28.13	51.08	105.17
Canola-soybean	-29.73	91.80	244.78
Canola-sunflower	-25.47	115.77	188.85
Canola-pearl millet	-22.37	11.77	65.05

Increased planting date flexibility makes buckwheat an acceptable alternative to double-crop soybean despite lower yield and return of buckwheat. As an emergency crop for late planting, buckwheat produces a positive net return after canola and wheat with average and optimum winter-crop yields, and after canola at base yields. Buckwheat has the potential to be used in a wide variety of food and feed products. The potential industrial uses for the crop have not yet been explored. Further expansion of the domestic and export market is necessary to ensure markets for increased production.

Pearl Millet

Double-crop pearl millet was the least profitable alternative except at the base yield level. Because costs of production for double-crop pearl millet are similar to those for double-crop soybean, the poor economic competitive ability of pearl millet is primarily due to low potential grain yields and price for the crop. Pearl millet probably needs additional plant breeding to produce consistent yields that would make it a competitive double-crop option. Market expansion for pearl millet is likely to occur, but at a relatively slow pace until further research makes the crop more viable. Pearl millet is particularly adapted to sandy soils and drought and may fit a need in Missouri under those conditions.

Sunflower

At average yields, double-crop sunflower had greater net returns than double-crop buckwheat, soybean and pearl millet after canola or wheat. Costs for sunflower were about 20 percent greater than for soybean, partly because of additional nitrogen and insecticide inputs for sunflower. Prices used here are those offered on contract for high-oleic acid sunflower varieties by regional processors in 1994. The price of sunflower and canola fluctuates relative to the price of other oilseeds, mainly soybean, with a corresponding flux in available markets for the crop. Some oilseed-type sunflower has also been used by regional birdseed processors. Increased industrial use of specialty sunflower varieties has opened new markets for the crop.

Other Double-crop Options

Besides the crops discussed here, a few other possible short-season crops could potentially be double cropped after wheat and canola. These include foxtail millet (Setaria italica), mung beans (Vigna raidata), cowpeas (Vigna unguiculata) and dry edible beans (Phaseolus spp.). Of these, dry edible beans have shown the greatest promise in Missouri.

Of the five double crops evaluated in the University of Missouri's two-year study, soybean had the lowest cost of production. At average yield levels, net returns were greatest for wheat-amaranth, canola-amaranth, wheat-sunflower and wheat-soybean systems. At average yields, the lowest net returns resulted from canola-pearl millet and canola-buckwheat systems. Canola was shown to be less profitable than wheat in double-cropping systems for central and southern Missouri. Sunflower was shown to be an agronomically and economically viable alternative to soybean following either canola or winter wheat.

Considering prices, available markets and agronomic quality, buckwheat and sunflower appear to be the most promising double-crop options when compared with soybean at average, base and optimum yields. Further breeding efforts and market expansion would improve the suitability of amaranth and pearl millet as double-crop options to soybean for the lower-Midwest region.

Disease Resistance in Fruit and Vegetables

Disease resistance in fruit and vegetables; There are a number of lines of defence against pests (that is, those animals that cause damage to the plants we grow) and diseases in the organic garden, principle among these being the practice of good husbandry, creating healthy soil and ensuring high standards of garden hygiene. But no matter how diverse and healthy the garden eco-system may be, there will always be a degree of disease and pest presence. In many ways, some level of pathogen population in the garden can be not only acceptable but desirable as they are indicative of a generally healthful and diverse environment, and add to the overall robustness of the system as an immunity to such detrimental influences will build up, particularly in a balanced polycultural regime. Indeed, most of the plants we grow will tend to be selected because they are trouble free, and those that are more susceptible to attack will have fallen by the wayside over time. However, most farmers find it unacceptable that the food crops they grow are damaged by pests.

For these crops there has been considerable research and selective breeding carried out in order to find cultivars that are resistant or immune to pest and disease damage. Generally this has involved finding suitable genetic material amongst existing stocks or in the wild, which is then incorporated into commercial varieties.

Example: The Apple

In the case of apples, in which research is being carried out in order to develop resistance to diseases such as apple scab (*Venturia inaequalis*), powdery mildew (*Podosphaera leucotricha*), orchard fireblight (*Erwinia amylovora*), woolly apple aphid (*Eriosoma lanigerum*) and collar rot (*Phytophthora cactorum*), the main sources of resistant material used in breeding programmes such as those being run by East Malling Research Station in England or Hortresearch in New Zealand are major gene resistances derived from crab-apples. The Vf gene for black spot resistance is derived from the ornamental crab-apple species *Malus floribunda*. Most black spot resistant cultivars developed around the world carry this gene, but there are some selections that carry the Vr (from M. pumila) or Vm (from M. micromalus) gene. Major gene resistances to powdery mildew are derived from M. robusta (Pl1) and M. zumi (Pl2), and the apple cultivar *Northern Spy* has a long-standing reputation for its major gene resistance to woolly apple aphid. Since early this century this resistance has been used to develop woolly aphid resistant rootstocks such as MM.106 and M.793. Much later it was shown that the cultivar was also very resistant to collar rot and a useful breeding parent for this resistance.

Resistance and Immunity

Some plants can tolerate the presence of large numbers of insects without being severely affected. This is not very satisfactory however as insects will still cause damage, and in fact further breeding and population expansion of the pest species is supported. Other varieties are less attractive to pests, but this can be difficult to sustain or demonstrate.

The most valuable form of resistance is where the pest cannot survive as well on one variety as on another. In some cases this can actually make the plants immune to attack, as is the case with the lettuces *Avoncrisp* and *Avondefiance* which were bred at the Institute of Horticultural Research, Wellsbourne during the 1960s, which are fully resistant to lettuce root aphid (*Pemphigus bursarius*).

Trade-off of Breeding for Resistance

Sometimes however there can be a *trade-off*, for those varieties which have increased immunity or resistance may be lacking in other qualities such as flavour, yield or quality. Celery resistant to the Fusarium fungus (*Fusarium oxysporum spp.*) may not succumb to this disease, but may also be unacceptably short, ribby and low yielding.

Further, a cultivar that is resistant to one disease may be more susceptible to another that is equally important. A lettuce cultivar that is resistant to mosaic virus may be sensitive to corky root disease (caused by *Rhizobius suberfaciens*), whilst another that resists corky root may be vulnerable to downy mildew (*Brim lactic*).

Another drawback to resistance is that depending on the host pathogen system, resistance is sometimes not long lasting as new pathogen strains quickly develop, and further research and breeding is constantly needed.

Availability of Resistant Varieties

Resistant varieties are not available for all crops. For several of the most damaging plant diseases, such as Potato blight (*Phytophthora infestans*) and white rot (*Sclerotic cepivorum*) of the Allium family, no acceptable resistant cultivars are yet available. In addition, commercial seed companies and plant breeders rarely invest resources into developing resistant cultivars for more minor or speciality crops, which often tend to be those of greater interest to the organic grower.

In general it is probably fair to say that resistance will not fully guarantee total crop protection, but choosing resistant varieties should rather be considered as a part of an overall Integrated pest management strategy, especially against virus diseases. In particular they can be especially useful where the threat from specific pests and diseases is high.

Table of Resistant Crops

The table below illustrates a number of pest and disease resistant vegetable crop varieties, although should not be considered comprehensive.

Crop	*Variety*	*Comments*
French Bean	Aigullion	Resistant to bean mosaic virus and anthracnose
	Hildora	Tolerant of bean mosaic virus
	Maxi	Tolerant of bean mosaic virus
	Nektar Queen	Resistant to bean mosaic virus
	Farba	Resistant to bean mosaic virus
	Hilda	Resistant to bean mosaic virus
Broad Bean	Futura	Tolerant of chocolate spot
Brussels sprouts	Cavalier	Resistant to light leaf spot
	Braveheart F1	Resistant to powdery mildew and light leaf spot
Cabbage	Stonehead F1	Resistant to mildew
Calabrese	Emperor F1	Highly tolerant of black rot and downy mildew
	Sampson F1	Highly resistant to black rot and downy mildew
Carrot	Fly Away F1	Bred for resistance to carrot fly
	Newmarket	Good resistance to splitting
Cucumber	Bush Champion F1	Resistant to cucumber mosaic virus
	Marketmore	Resistant to cucumber mosaic virus
	Burpless Tasty Green F1	Resistant to mildew and tolerant of heat
	Slice King F1	Resistant to downy and powdery mildew, gummosis and angular leaf spot
	Cumlaude F1	Tolerant of powdery mildew
	Media F1	Tolerant of powdery mildew
Leek	Swiss Giant, Zermatt	Moderate rust resistance
	Alvitta	Rust Tolerant
	Conora	Good resistance to rust
Lettuce	Alexandria	Highly resistant to downy mildew and bolting
	Lisbusa	Resistant to downy mildew, tolerant of mosaic virus, root aphid and tipburn
	Sylvestra	Resistant to downy mildew and aphids
	Little Gem	Resistant to root aphid
	Barcelona	Resistant to aphids and mildew

	Iglo	Resistant to downey mildew
	Revolution	Resistant to mildew and bolting
	Smile	Resistant to aphids, mildew, tipburn and bolting
Parsnip	Avonreister	Very good rsistance to canker
	Gladiator F1	High resistance to canker
	Tender and True	Good canker resistance
Pea	Ambassador	Resistant to powdery mildew, entation virus and fusarium wilt
	Cavalier	Good resistance to powdery mildew.
	Greenshaft	Resistant to downy mildew and fusarium wilt
	Rondo	Resistant to fusarium wilt
Potato (first early)	Colleen	Good resistance to blight and scab
	Premiere	Resistant to blight and spraing
	Pentland Javelin	Good all round disease resistance, including golden eelworm and scab
	Swift	Resistant to golden eelworm and tolerant to blackleg
Potato (second early)	Cosmos	Resistant to blight and common scab
	Osprey	Very resistant to scab and eelworm
Potato (maincrop)	Harmony	Very resistant to scab and partially resistant to white and golden eelworm
	Milva	Blight resitant
	Remarka	Good all around disease resitance
	Admiral	Good resistance to blight and scab
	Cara	Blight resitant
	Valor	Good overall disease resistance combined with high eelworm and blight resistance
Pepper	Bell Boy F1	Mosaic virus resistant
	Mandy F1	Highly resistant to mosaic virus
Sweetcorn	Golden sweet F1	Highly resistant to smut
Tomato	Alicante	Good resistance to greenback and mildew
	Libra F1	Resistant to fusarium crown and root rot
	Shirley F1	Resistant to fusarium, cladosporium and TMV

Agricultural Experiment Station

An agricultural experiment station is a research center that conducts scientific investigations to solve problems and suggest improvements in the food and agriculture industry. Experiment station scientists work with farmers, ranchers, suppliers, processors, and others involved in food production and agriculture. They have made outstanding contributions to the development of food and agriculture in the United States and Canada.

Location of Stations

The United States has more than 50 stations (1988), run by about 13,000 scientists (1988). Each state has at least one main station, usually located at and associated with a land-grant university. Many states have branch stations to meet the special needs of different climate and geographical zones in those states.

The US Department of Agriculture also directly maintains several experiment stations, including the Henry A. Wallace Beltsville Agricultural Research Center in Maryland. The Beltsville station contains the main building of the National Agricultural Library. The U.S. National Arboretum in Washington, DC is a division of the Beltsville Agricultural Research Center.

Funding

The U.S. experiment stations are state institutions. However, the federal and state governments cooperate in funding the research done at the stations. The states provide about 60 per cent (1988) of the government money. Additional income comes from grants, contracts, and the sale of products. The stations receive a total income of more than $1 billion a year.

Research

Station scientists study a biological, economic, and social problems of food and agriculture and related industries in each state. They investigate such areas as crop variations, soil testing, livestock, processing and animal technology, and other advanced technology to food and agriculture. They also work with specialists called extension agents. These specialists help inform farmers about developments in agriculture. Most agricultural experiment station scientists are faculty members of the land-grant universities.

History

The first state agricultural experiment station in the United States was organised in 1875 at Wesleyan University in Middletown, Connecticut. It was supported by private donations and state funds. Organization and method were modeled upon European stations and the work of state chemists. By 1887 fourteen states had definite organizations and in thirteen others the colleges conducted equivalent work. The Bussey Institution at Harvard University (since 1871) and the Houghton Farm at Cornwall, N. Y. (1876-88), were privately endowed stations. Federal aid for state experiment stations began with the Hatch Act of 1887. The Hatch Act authorized direct payment of federal grant funds to each state to establish an agricultural experiment station for its land-grant college. Land-grant colleges had been established under the Morrill Act of 1862. The aid was increased by the Adams Act (1906) and the Purnell Act (1925). The provisions of the original Hatch Act and of later legislation providing increasing funds were combined in the Hatch Act of 1955.

One of the oldest agricultural experiment stations is the Rothamsted Experimental Station, located at Harpenden in Hertfordshire, England, where the great statistical geneticist Ronald Fisher was inspired to important advances in the theory of statistical inferences and genetics.

Recent history

The McIntire-Stennis Act of 1962 authorized forestry research studies at experiment stations. The federal government takes part in the experiment station program through the Cooperative State Research Service of the U.S. Department of Agriculture. The department coordinates research activities among the state stations.

Canada

In Canada, about 50 per cent (1988) of the experiment stations are controlled by the Canadian government. The Central Experimental Farm in Ottawa is the headquarters of the federal system. Private industries, universities, and agricultural colleges control the remainder of the stations. Each province has a number of provincial stations.

Good Agricultural Practices

The term Good Agricultural Practices can refer to any collection of specific methods, which when applied to agriculture, produce results that are in harmony with the values of the proponents of those practices. There are numerous competing definitions of what methods constitute "Good Agricultural Practices", so whether a practice can be considered "good" will depend on the standards you are applying.

The remainder of this article describes one particular definition of "Good Agricultural Practices" as defined by the Food and Agriculture Organization (FAO) of the United Nations.

Description of the UN FAO GAPs

Good Agricultural Practices (or GAPs) are a collection of principles to apply for on-farm production and post-production processes, resulting in safe and healthy food and non-food agricultural products, while taking into account economical, social and environmental sustainability.

Though the term is not new, it has really begun to get wide attention at the end of the 1990s.

GAPs may be applied to a wide range of farming systems and at different scales. They are applied through sustainable agricultural methods, such as integrated pest management, integrated fertilizer management and conservation agriculture. They rely on four principles:

- Economically and efficiently produce sufficient (food security), safe (food safety) and nutritious food (food quality);
- Sustain and enhance natural resources;
- Maintain viable farming enterprises and contribute to sustainable livelihoods;
- Meet cultural and social demands of society.

The concept of GAPs has changed in recent years because of a rapidly changing agriculture, globalization of world trade, food crisis (mad cow disease), nitrate pollution of water, appearance of pesticide resistance, soil erosion...

GAPs applications are being developed by governments, NGOs and private sector to meet farmers and transformers needs and specific requirements. However, many think these applications are only rarely made in a holistic or coordinated way.

They provide the opportunity to assess and decide on which farming practices to follow at each step in the production process. For each agricultural production system, they aim at allowing a comprehensive management strategy, providing for the capability for tactical adjustments in response to changes. The implementation of such a management strategy requires knowing, understanding, planning, measuring, monitoring, and record-keeping at each step of the production process. Adoption of GAPs may result in higher production, transformation and marketing costs, hence finally higher costs for the consumer.

GAPs require maintaining a common database on integrated production techniques for each of the major agro-ecological area, thus to collect, analyze and disseminate information of good practices in relevant geographical contexts.

Good Agricultural Practices Related to Soil

- Reducing erosion by wind and water through hedging and ditching
- Application of fertilizers at appropriate moments and in adequate doses (i.e., when the plant needs the fertilizer), to avoid run-off.

- Maintaining or restoring soil organic content, by manure application, use of grazing, crop rotation
- Reduce soil compaction issues (by avoiding using heavy mechanical devices)
- Maintain soil structure, by limiting heavy tillage practices

Good Agricultural Practices Related to Water

- Practice schedule irrigation, with monitoring of plant needs, and soil water reserve status to avoid water loss by drainage
- Prevent soil salinization by limiting water input to needs, and recycling water whenever possible
- Avoid crops with high water requirements in a low availability region
- Avoid drainage and fertilizer run-off
- Maintain permanent soil covering, in particular in winter to avoid nitrogen run-of
- Manage carefully water table, by limiting heavy output of water
- Restore or maintain wetlands.
- Provide good water points for livestock

GAPs Related to Animal Production, Health and Welfare

- Respect of animal well-being (freedom from hunger and thirst; freedom from discomfort; freedom from pain, injury or disease; freedom to express normal behavior; and freedom from fear and distress)
- Avoid nontherapeutic mutilations, surgical or invasive procedures, such as tail docking and debeaking;
- Avoid negative impacts on landscape, environment and life: contamination of land for grazing, food, water and air
- Check stocks and flows, maintain structure of systems
- Prevent chemical and medical residues from entering the food chain
- Minimize non-therapeutic use of antibiotics or hormones
- Avoid feeding animals with animal wastes or animal matter (reducing the risk of alien viral or transgenic genes, or prions such as mad cow disease),
- Minimize transport of live animals (by foot, rail or road) (reducing the risk of epidemics, e.g., foot and mouth disease)
- Prevent waste run-off (e.g. nitrate contamination of water tables from pigs), nutrient loss and greenhouse gas emissions (methane from cows)

- Prefer safety measures standards in manipulation of equipment
- Apply traceability processes on the whole production chain (breeding, feed, medical treatment...) for consumer security and feedback possibility in case of a food crisis (e.g., dioxin).

Modernisation in Agriculture

During the latter half of the twentieth century, what is known today as modern agriculture was very successful in meeting a growing demand for food by the world's population. Yields of primary crops such as rice and wheat increased dramatically, the price of food declined, the rate of increase in crop yields generally kept pace with population growth, and the number of people who consistently go hungry was slightly reduced. This boost in food production has been due mainly to scientific advances and new technologies, including the development of new crop varieties, the use of pesticides and fertilizers, and the construction of large irrigation systems.

Basic Practices of Modern Agricultural Systems

Modern agricultural systems have been developed with two related goals in mind: to obtain the highest yields possible and to get the highest economic profit possible. In pursuit of these goals, six basic practices have come to form the backbone of production: intensive tillage, monoculture, application of inorganic fertilizer, irrigation, chemical pest control, and genetic manipulation of crop plants. Each practice is used for its individual contribution to productivity, but when they are all combined in a farming system each depends on the others and reinforces the need for using the others. The work of agronomists, specialists in agricultural production, has been key to the development of these practices.

Intensive Tillage

The soil is cultivated deeply, completely, and regularly in most modern agricultural systems, and a vast array of tractors and farm implements have been developed to facilitate this practice. The soil is loosened, water drains better, roots grow faster, and seeds can be planted more easily. Cultivation is also used to control weeds and work dead plant matter into the soil.

Monoculture

When one crop is grown alone in a field, it is called a monoculture. Monoculture makes it easier to cultivate, sow seed, control weeds, and harvest, as well as expand the size of the farm operation and improve aspects of profitability and cost. At the same time, monocultures tend to promote the use of the other five basic practices of modern agriculture.

Use of Synthetic Fertilizers

Very dramatic yield increases occur with the application of synthetic chemical fertilizers. Relatively easy to manufacture or mine, to transport, and to apply, fertilizer use has increased from five to ten times what it was at the end of World War II (1939-45). Applied in either liquid or granular form, fertilizer can supply crops with readily available and uniform amounts of several essential plant nutrients.

Irrigation Technologies

By supplying water to crops during times of dry weather or in places of the world where natural rainfall is not sufficient for growing most crops, irrigation has greatly boosted the food supply. Drawing water from underground wells, building reservoirs and distribution canals, and diverting rivers have improved yields and increased the area of available farm land. Special sprinklers, pumps, and drip systems have greatly improved the efficiency of water application as well.

Chemical Pest Control

In the large monoculture fields of much of modern agriculture, pests include such organisms as insects that eat plants, weeds that interfere with crop growth, and diseases that slow plant and animal development or even cause death. When used properly, synthetic chemicals have provided an effective, relatively easy way to provide such control. Chemical sprays can quickly respond to pest outbreaks.

Genetic Manipulation

Farmers have been choosing among crop plants and animals for specific characteristics for thousands of years. But modern agriculture has taken advantage of several more recent crop breeding techniques. The development of hybrid seed, where two or more strains of a crop are combined to produce a more productive offspring, has been one of the most significant strategies. Genetic engineering has begun to develop molecular techniques that selectively introduce genetic information from one organism to another, often times from very unrelated organisms, with a goal of capitalizing on specific useful traits.

But for almost every benefit of modern agriculture, there are usually problems. Excessive tillage led to soil degradation, the loss of organic matter, soil erosion by water and wind, and soil compaction. Large monocultures are especially prone to devastating pest outbreaks that often occur when pests encounter a large, uniform area of one crop species, requiring the continued and excessive use of chemical sprays. When used excessively, chemical fertilizers can be easily leached out of the soil into nearby streams and lakes, or even down into underground water supplies. Farmers can become dependent on chemical pest and weed control. Modern farm systems lack the natural control agents needed for biological pest management, and larger amounts of sprays must be used as pests rapidly evolve resistance. People also worry about chemical pollution of the environment by sprays and fertilizers, and the possible contamination of food supplies. Modern agriculture has become such a large user of water resources that overuse, depletion, saltwater contamination, salt buildup in soil, fertilizer leaching, and soil erosion have become all too common. Agricultural water users compete with urban and industrial use, and wildlife as well. Hybrid seed has contributed greatly to the loss of genetic diversity and increased risk of massive crop failure, as well as an increased dependence on synthetic and non-renewable inputs needed for maintaining high yield. Genetically engineered crops have the same negative potential, especially as the selection process takes place less and less in the hands of farmers working in their own fields, but rather in far away laboratories. In the future, in order to take advantage of new technologies and practices, farming systems will need to be viewed as ecosystems, or agricultural ecosystems. By monitoring both the positive and negative impacts of modern farming practices, ecologically based alternatives can be developed that protect the health of the soil, air, and water on farms and nearby areas, lower the economic costs of production, and promote viable farming communities around the world. Organic agriculture, conservation tillage, integrated pest management (IPM), and the use of appropriate genetic techniques that enhance local adaptation and variety performance are a few of the possible ways of ensuring the sustainability of future generations of farmers.

Plan for Modernisation of Agriculture

The PMA is part of Uganda's Poverty Eradication Action Plan (PEAP). It aims at improving incomes, reducing food insecurity, creating gainful employment and a good environment for sustainable natural resource management. The PMA provides the principles and priority intervention programmes and modalities for implementing holistic and multi-sectoral programmes that have greater opportunities of improving the livelihoods of the agricultural based sections of the population through better understanding of the opportunities available to specific communities, strengthening participatory planning processes and improving the quality of service delivery. The PMA is neither a project nor a programme. It provides the principles and framework for the design and implementation of programmes and projects that impact on the agricultural based livelihoods.

The Vision

Agricultural transformation in Uganda will lead to *"poverty eradication through a profitable, competitive, sustainable and dynamic agricultural and agro-industrial sector"*. Achieving this vision will depend on two related processes: transforming the subsistence farmer, and transforming the agricultural sector in general. The transformed subsistence farmer is envisaged to be one that:

- is producing and selling more to the market without compromising household food security;

- has access to information on market prices and is well linked to both product and input markets;
- lives on a well-planned and managed farm, has record keeping skills and is willing to adopt new proven technologies;
- manages farm activities in a manner that does not degrade the environment.

The transformation of the subsistence farmer, together with the macroeconomic and sectoral policies, the institutional reforms and adjustments, and the PMA interventions will result in an agricultural sector that is:

- competitive (with lower per unit cost of production, processing and marketing);
- transformed (using high-yielding, disease- and pest-resistant planting materials, fertilizers, pesticide, and high-yielding livestock and poultry breeds);
- diversified (producing high-value commodities and higher income-elasticity of demand commodities especially fruits, vegetables, fish and livestock products);
- export oriented (with increased trade in traditional cereals into regional markets and food contracts; and increased traditional exports of coffee, tea, cotton and tobacco, but in valueadded forms rather than as raw materials);
- less vulnerable to food insecurity by balancing the supply and demand side of the food security equation;
- effectively utilizing resources such as land, water and forests in a sustainable manner for both the present and future generations.

Objectives

The PMA is a poverty focused framework of principles whose main objectives are to:

- increase incomes and improve the quality of life of poor subsistence farmers through increased productivity and increased share of marketed production;
- improve household food security through the market rather than emphasizing self sufficiency;
- provide gainful employment through the secondary benefits of PMA implementation such as agro-processing factories and services;
- promote sustainable use and management of natural resources by developing a land use and management policy and promotion of environmentally friendly technologies.

The Missio

In order to achieve the vision and its objectives, the mission of the PMA is eradicating poverty by *"transforming subsistence agriculture to commercial agriculture"*. This mission derives from the PMA's goal of poverty eradication through agricultural transformation and sustainable natural resource –based livelihood.. Improving the welfare of poor subsistence farmers will require that they re-orient their production towards the market. More of their production must be marketed to enable them to earn higher incomes. In a broad sense, achieving the vision will engender maintaining prudent macroeconomic and sectoral policies, undertaking the institutional reforms and adjustments, and implementing the interventions.

Strategies

Successful implementation of the PMA will build on the following strategies:

- Making poverty eradication the overriding objective of agricultural development;
- Deepening decentralization to lower levels of Local Governments for efficient service delivery;
- Removing direct involvement of government in commercial aspects of agriculture and promoting the role of the private sector;
- Supporting the dissemination adoption of productive enhancing technologies.
- Guaranteeing food security through the market and improved incomes, thereby allowing households to specialize, rather than through household self sufficiency;
- Ensuring that all intervention programmes are gender focused and gender responsive;
- Promoting a two-way planning and budgeting process by empowering local governments and enabling them to influence public policy and allocate resources to alleviate location specific constraints in a non-sector manner;
- Ensuring coordination of the multi-sector interventions to remove any constraints to agricultural modernisation.

The interventions of the PMA seek to increase the productivity of factors of production in agriculture, to ensure food security, to create gainful employment, to increase incomes, and to improve the quality of life of those engaged in the agriculture sector. The PMA identifies seven areas of intervention into which public investments will be built.

PMA Priority Areas

- Agricultural research and technology development
- Delivery of agricultural advisory services
- Rural financial services
- Promotion of agro processing and agricultural marketing
- Agricultural education
- Natural resource management
- Physical infrastructure

Six Fallacies of Modern Agriculture

Farm subsidies are over $20 billion a year, farmers go broke anyway, pesticides contaminate water and food, and topsoil washes down the river. Why is that year after year, under Republicans and Democrats, our agriculture system has persistent troubles?

Wendell Berry, a Kentucky farmer and a writer, says it's because we operate from assumptions about agriculture that are just plain wrong. In his new book Home Economics, he lists six fallacies of modern agriculture — fallacies that most people believe deeply, especially in the High Places where farm bills are generated.

Here they are. Judge for yourself.

1. Agriculture may be understood and dealt with as an industry.

Industry deals with dead materials, mechanical processes, cold rationality, factories that wear out, workplaces that are distinct from places to live. Agriculture deals with living things, biological processes, nurture and care, farms that should last for generations, and workplaces that are also homes.

Treating a farm like a factory gives it the short life expectancy, the unlivability, the exploitive nature of the factory. The U.S. counties with the largest, most industrialized farms also have the highest rural levels of chemical contamination and of poverty. That's no surprise, says Wendell Berry.

2. A sound agricultural economy can be based on an export market.

The 1985 Farm Bill was based on one simple idea — that our farmers have to compete on the international market. But Berry says a sound economy cannot be based on a market beyond its control. For stability, security, quality control, freshness, and low transport costs, every nation should produce food primarily for itself.

The farm should first supply the needs of the farm family, then the needs of the surrounding community. Distant trade should be in occasional surplus only. Surplus should be global insurance against emergency, not something for one nation to count on exporting and another on importing.

A subsistence-based farm economy, Berry says, would give us a healthier food supply, a more diverse local agriculture, more local employment, and less vulnerability to the perturbations of global trade.

3. The "free market" can preserve agriculture.

The market sets a price for a farm product, but not for the sources of that product — the topsoil, the ecosystem, the farm family, the farm community. It treats those sources as free or expendable. Berry says, "If a squash on the table is worth more than a squash in the field, and a squash in the field is worth more than a bushel of soil, that does not mean that food is more valuable than soil. It means that we do not know how to value the soil."

To lean back and expect a mystical mechanism called the market to lead to proper decisions is, says Berry, either lazy or felonious. Economics should come from human values, those values that can be set by the market and those that transcend the market. If farmers can't "afford" to take soil conservation measures or cut use of dangerous chemicals or stop overdrafts of groundwater, if we can't "afford" to keep farmers in farming, then we need some larger definition of the word "afford" than the market can give.

4. Productivity is the measure of agricultural success.

Berry is not impressed that the American farmer feeds "57 or 75 or God knows how many people". What are the costs of that productivity, he asks, in soil, farms, and farmers, in pollution of water and food, in the vulnerability of the food supply?

To celebrate output without questioning input is like rejoicing in the water flowing out of the dam without wondering whether any water is flowing in to replace it. We should celebrate thrift, thriving, health, says Berry, not sheer output.

5. There are too many farmers.

Agricultural economists, not farmers, keep saying there are too many farmers. They never say that there are too many agricultural economists.

There is too much production in the United States, but that does not mean there are too many farmers. There are too few farmers to take proper care of the land. Farmers are being replaced by machines and chemicals not because there are too many farmers, but because we have chosen not to pay for food what it really costs, and not to pay farmers as much as agricultural economists or combine manufacturers. That is a social decision, not a law of nature.

6. Hand labor is bad.

The goodness of any labor depends on its scale, on its compensation, and on who is doing it for whom. Americans happily do hand labor in their gardens for no more compensation than fresh vegetables and praise from the family. Helping plants to grow is inherently good work, unless it's done acre upon acre, day after day, in someone else's field, at poverty wages.

Berry says, "any work is miserable, whether done by hand or by machine, if it is economically desperate."

To my knowledge no one in decades has stood up in Congress and talked about farming as something to be done by more people on smaller farms with less mechanization and more care, producing less output of higher quality at less cost for greater profit. Maybe that's a kind of farming that would work better than what we've got.

4

Seed Technology and Plant Growth: Concepts & Approaches

New Seed Technology Threatens Farmers

New agro-technology that genetically-disables a seed's capacity to germinate could threaten the livlihood of traditional farmers, mainly in developing countries, say researchers. The U.S. Department of Agriculture (USDA) and a large American cotton seed company, have patented a jointly-developed technique that enables seed companies to switch a plant's reproductive processes on and off. This means that if farmers attempt to replant harvested seed, it would be sterile. Now the concern among farmers is that they will have to spend much more of their money to make annual seed purchases instead of saving seed and depending upon their own plant breeding skills as they have done for centuries.

"It's terribly dangerous," says Hope Shand, research director of the North-American based Rural Advancement Foundation International (RAFI). "Half the world's farmers are poor and depend upon saved seed and their own breeding skills because they can't afford to buy seed every growing season. Yet, poor farmers grow 15 to 20 percent of the world's food and they directly feed at least 1.4 billion people — 100 million in Latin America, 300 million in Africa, and one billion in Asia.

"It is a crime that U.S. tax payer's money went towards developing this new technology."

Mississippi-based Delta and Pine Land, the largest cotton seed corporation in the United States, argues that this new technology — which may be applicable to all crops in the future — will lead to more research on staple foods grown in developing countries that have been difficult to turn into hybrid seeds, such as rice and wheat.

"The patent has the prospect of opening significant worldwide seed markets to (biologically engineered) crops, in which seed currently is saved and used in the next season," according to a statement by Delta and Pine Land.

Currently, private companies are discouraged from investing in crops that are not easily 'hybridised', such as wheat and rice. Hybrid seeds, such as corn, are attractive to corporations because they yield a higher profit as they do not reproduce themselves and farmers have to buy new supplies every year.

While farmers have been unable to save hybrid seeds, they have continued to save seeds from traditional non-hybridised crops. But that may all change with this new patented technology.

Crops that are difficult to hybridise — generally ignored by multinational seed corporations because of the low profit potential — now may get more attention.

So far, only tobacco seeds have responded to the new technique, but Delta and Pine Land plans to apply this technology to rice, wheat, sorghum and soybeans after the year 2000, primarily targeting markets in developing countries.

USDA spokesman Michael Ruff acknowledges the downside for farmers as they may be forced to pay more for seed stocks every year but says it is too early to tell if this new technology will be positive or negative for farmers. "It's really all speculation," he told IPS. "Until further research is done, we really won't know if this new technology can even be applied to other crops."

But he admits that, if the new technique can be applied to crops such as wheat and rice, private companies will be encouraged to invest in crops typically unable to hybridise, so that they could make higher profits on crops that must be purchased each year.

Delta and Pine Land maintain that the new research interest in the "forgotten" crops will lead to a choice of more higher-yielding plant varieties for farmers and, therefore, result in greater global food security. But farmers and seed researchers have strong doubts about this "miracle" technology.

"This is the neutron bomb of agriculture," says Camila Montecinos, an agronomist with the Chilean-based Center for Education and Technology, who is calling for a global boycott of the new technology. "This is an immoral technique that robs farming communities of their age-old right to save seed and their role as plant breeders."

Non-governmental organizations fear that public sector plant breeders wanting access to patented seeds will be forced to adopt the new technology as a licensing requirement, for example. As better off farmers would be forced to pay for the new seed technology, their poorer neighbours would no longer be able to exchange breeding material with their fellow farmers, as has been done traditionally.

Neth Dano, director of the Philippines-based SEARICE, an organisation that works with farmers in South-east Asia, says poor farmers traditionally select the best seeds every year by crossing commercial varieties with other rice strains and over time this locally-adapted rice molds itself to the farm's distinct ecosystem.

"This new technology could put an end to all this and increase crop uniformity and vulnerability. Far from improving plant breeding, the new technology could drive hundreds of millions of farmers out of plant breeding and, since no one else will breed for their needs, out of agriculture altogether," she says. Researchers also worry that even if farmers do not buy the new seed, if nearby fields are planted with this new seed technology they could find that the following season some of their seeds will be sterile, adds Montecinos. "Crops geneticists who have studied the patent say it is likely that pollen from crops carrying the new trait will infect the fields of farmers who either reject or can't afford the technology. When farmers reach into their bins to sow seed the following season they could discover — too late — that some of their seed is sterile."

If true, she says, this could lead to huge losses in crop yields and irregular harvests for several years and a general decline in food security for the poorest farm communities.

Description of the Seed Division's activities and resources

Description of The Seed Science and Technology Division

Objectives

- General: help promote and develop the local seed industry using the sustainable agriculture framework through seed education, research and extension
- Specific 1: make the general public, seed producers, farmers, institutions and organizations (both formal and informal) aware of the value of use of quality and appropriate seed, locally developed (indigenous) seed practices and farming systems
- Specific 2: provide a venue for dissemination and exchange of information on seed sources, availability, technology and related matters among practitioners, development workers, Gos, NGOs and People's Organizations (POs)
- Specific 3: encourage farmer-based or community-based seed production and genetic conservation endeavors
- Specific 4: generate data and produce information/learning materials to answer seed-related information gaps through primary and secondary research

Activities

- Collect, disseminate and exchange information related to seed, genetic resources and biodiversity, agroforestry, sustainable agriculture (through correspondence, visits, seminars, etc.);

- Maintain relevant databases on 1) seed sources and availability, 2) seed production/handling technologies, 3) seed/plant characteristics 4) other seed-related information develop databases and produce extension materials/ publications on seeds and related issues;
- Conduct practical and basic research on seed production/technology and genetic conservations conduct training at the university and on-site for farmers, GOs, NGOs and Pos; also offer a yearly short course (summer) on seed production, handling and genetic conservation using sustainable agriculture as framework;
- Maintain a learning resource facility (which contains references, videos, posters, slides, etc.);
- Network with different sectors involved in seed matters.

Extension Materials / publications

1. Agroforestry Seeds. Sustainable Agriculture Newsletter vol.2 no.3 (1990); Supplement on Agroforestry Seed Sources; Index on Agroforestry Species.
2. Agroforestry Seeds Circular no.1 (1991), no.2 (1992) and no.3 (1993).
3. Supplement to Agroforestry Seeds Circular no.1. Philippine agroforestry species scientific name, official, common and local names, uses, sources/ location; Index of Agroforestry Seeds Sources (Philippines).
4. Supplement to Agroforestry Seeds Circular no. 2. Philippine agroforestry species scientific name, official, common and local names, uses, sources/ location and uses; Index of Agroforestry Seeds Sources (Phil.).
5. Supplement to Agroforestry Seeds Circular no.3. (Agroforestry Species in the Philippines: Common Names, Scientific Name and Growth Requirement; Index of Agroforestry Seeds Sources (Philippines).
6. Some chapters in the Resource Book on Sustainable Agriculture in the Uplands. 1990. IIRR, MBRLC and Mag-uugmad Foundation, Inc. Philippines.
7. Some chapters in the Low-external Input Rice Production Kit, 1990. IIRR, Philippines.
8. Manwal sa Pagpaparami at Pangangalaga ng Binhi (Manual on Seed Production and Handling). 1992. Joint publication of the Department of Agronomy, UP Los Baos and International Institute of Rural Reconstruction (IIRR), Silang, Cavite, Philippines.
9. Seeds and Plant Propagation. 1992. Agroforestry Technology Information Kit No. 5. IIRR, Silang, Cavite, Philippines.

10. Discussion/conference papers on seeds and sustainable agriculture.
11. Training Aid/Manual on Sustainable Seed Production/Technology. 1994.
12. Intellectual Property Rights (IPR) Sourcebook Philippines With Emphasis on Intellectual Property Rights in Agriculture and Food. UPLB and MODE, Philippines. 1994.
13. Soundslide on "sustainable" seed production, technology and genetic conservation (in Filipino).

Databases

14. The "Dirty Dozen" Pesticides.
15. Storage and Longevity of Recalcitrant Seeds.
16. Asexual Propagation Techniques.
17. Agricultural Crop Seed Production Data.
18. Natural Crop and Seed Protection Substances, Techniques/Practices.
19. Agroforestry Pests and Diseases in the Philippines.
20. Glossary of Terms Related to Agroforestry, Development, Genetic Conservation, Indigenous Knowledge System, Seed Production Handling Technology, Sustainable Agriculture & related fields/issues.
21. Indigenous Seed Practices.
22. Agencies/organizations/individuals doing seed-related work (seed education extension, research, production, conservation/seed banking, exchange, trading, etc.).
23. Pregermination Treatments of Selected Agroforestry Species.
24. Rapid Viability Testing Using Tetrazolium for Some Agroforestry Species.
25. Abstract/Bibliography of seed technology research in the Philippines.

Seed Information and Learning Resources and Services

The AFSICH (Agroforestry Seed Information Clearinghouse) Project ended in June 1994. However, the resources that have been developed or collected through the project will continue to be available to the network. The facilities are maintained by the Seed Science and Technology Division, Department of Agronomy. Activities of the Division is supported by the Seed Research and Development Project of UPLB. Copies of some learning materials can be made available, but a minimum fee will be asked to cover the cost of printing/copying, handling and mailing. Below are some of these resources and services:

Resources

1. Seed production and handling manual (Pagpaparami at Pangangalaga ng

Binhi). 1992. A collaborative effort between the Department of Agronomy, University of the Philippines Los Banos (UPLB) and the International Institute of Rural Reconstruction (IIRR). 34 pp. (available only in Tagalog; richly illustrated; based on sustainable agriculture framework; translation to appropriate languages, especially by NGOs or POs, is welcome).

2. "Agroforestry Seeds" issue, Sustainable Agriculture Newsletter 2(3), 1990. A collaborative effort of CUSO, IDRC, IIRR, SIBAT, UPLB through AFSICH and CADI. 36 pp. Includes topics on farm level technologies on various aspects of seed production and handling; a list of resource persons in sustainable agriculture, as well as agencies/ organizations, projects doing agroforestry seed-related activities; includes a supplement on local and international agroforestry seed sources.

3. Agroforestry Seeds Circular and Supplement. (AFSICH, Department of Agronomy, UPLB).

- Issue 1, 1991 (10 pp.). Includes seed research results; training notes; reader feedback; references and a calendar of activities on seeds, agroforestry and sustainable agriculture. The supplement (26 pp.) lists scientific name, common names and seed sources of 276 Philippine agroforestry species; also contains an index of Philippine agroforestry seed sources.
- Issue 2, 1992 (24 pp.). Features an NGO-farmer based seed collection and reforestation using indigenous species in Mindanao, Philippines; includes topics or short articles on quality of seeds from various sources; upgrading seed quali-ty through invigoration; pests and diseases in agroforestry; seed-related matters in multipurpose cover crops and green manure species; training notes; tips on handling seeds for transport; also includes reader feedback, useful references and calendar of activities on seeds, agroforestry and sustainable agriculture. The supplement (43 pp.) lists 385 agroforestry species in the Philippines with their official name, common names, seed sources and uses; also includes index of agroforestry seed sources (Philippines).
- Issue 3, 1993 (60 pp.). Features an article on entrusting seed production to farmers; also in-cludes topics or articles on seed germination of indigenous Philippine trees; seeds and fruit as medicine; recalcitrant seeds and intermediates; training and network notes; agroforestry seed procurement tips; endangered Philippine forestry and agroforestry species; species identification; and reader feedback of previous issues; also includes useful references and a calendar of activities on seeds, agroforestry and sustainable agriculture. The supplement (24 pp.) lists agroforestry species in the Philippines classified according to family and provided with common names, scientific name and growth requirements; also includes index of agroforestry seed sources (Philippines).

4. "Seeds and Plant Propagation", 106 pp. Booklet No. 5 of the Agroforestry Technology Information Kit (ATIK) 1992. This is a revision of a portion of the ATIK (published in 1989 by IIRR, Ford Foundation, and DENR in cooperation with UPLB) and of the Agroforestry Seeds Handling, 1990 (AFSICH, Department of Agronomy, UPLB).

5. Sound-slides on "sustainable seed production and handling" (in Tagalog/ Filipino; comes with a photo-text guide booklet which may be obtained separately). 1992.

6. IPR Sourcebook Philippines With Emphasis on Intellectual Property Rights in Agriculture and Food. 1994. UPLB and MODE. College, Laguna, Philippines. 173 pp. This is a compilation of materials that were presented during a conference on IPR issues in GATT, held last August at the Agricultural College of the University of the Philippines, Los Baos, (UPLB). Scientists, farmers, NGOs and government came together during the conference to exchange views on how IPRs on life will affect Philippine society. The papers included discussions on problems created by the current push to patent life from all sorts of angles: legislative, transnational corporations and trade, research, seed conservation, ethics, and international perspectives. It illustrates the hot debate going on in the Philippines.

7. Training aid/manual on sustainable seed production/technology (in Tagalog/ Filipino). 1994. Presents seed production and sustainable agriculture principles through questions which are based mostly on illustrations. Also included are an answer guide and instructions for the preparation of various set ups for demonstration and practical exercises. Active discussion and sharing are encouraged through group work.

8. Glossary of terms related to agroforestry, seeds, sustainable agriculture, genetic resources, etc.

9. Books, articles, theses, magazines, video tapes. Topics include seed biology, seed production technology, quality testing, nursery management, alternative crop production and protection systems (including indigenous systems), agroforestry, genetic conservation, sustainable agriculture, current issues, etc.

10. Posters on seeds, genetic diversity, and environmental issues (obtained from various sources; useful for training and formal instruction).

11. Documentation of training/short courses on seed technology/seed production, nursery management and sustainable agriculture conducted by the staff.

12. Database/listing on:

- Agroforestry seed sources, local and some international traders or organizations (species, prices and volume available are not necessarily updated and may be provided only under certain situations)

- Agencies, organizations, individuals doing seed-related work (seed education, extension, research, production, conservation/seedbanking, exchange, trading, etc.)
- Agroforestry pests and diseases in the Philippines
- Botanical pesticides and other natural means of crop/seed protection and pest management
- Indigenous seed/crop production and protection practices
- Seed storability of agroforestry, recalcitrant and agricultural crops species
- Multiplication/production techniques of asexually propagated crops
- Agronomic data for seed production of agricultural crops, including a number of indigenous vegetables
- Pregermination treatments of selected agro-forestry species
- Rapid viability testing using tetrazolium for some agroforestry species
- Recalcitrant seed storage techniques

13. Compendium of dangerous or banned pesticides and their ill effects

14. Seed herbarium of some indigenous agroforestry, forestry and agricultural species including a number of indigenous rice seeds.

Services

1. Conduct a training on seed production handling-technology, nursery management, genetic conservation, sustainable agriculture and community seed-banking upon request.
2. Conduct a regular two-week short course on "Seed Production, Handling and Genetic Conservation for Sustainable Agriculture"
3. Provide information/sourcing on seed availability, seed technology, agroforestry, sustainable agriculture and other related topics.
4. Provide assistance to farmers, POs and NGOs in setting-up seed production or seed banking programs.
5. Conduct (to a limited extent) seed quality tests for the net-work.
6. Conduct practical research on some seed aspect as identified by farmers/ practitioners or rural development workers.

Importance of Seed discolouration in Faba Beans *(Vicia fabae)* Grown in Southern Australia

Ascochyta blight, caused by *Ascochyta fabae* Speg. and chocolate spot caused by

Botrytis fabae Sardina and *B. cinerea* Fr. are the two most important diseases of faba bean (*Vicia fabae* L.) in southern Australia. All of the cultivars currently grown in Australia are susceptible to at least one of these diseases. Both diseases can cause significant crop losses and discolouration of grain, which reduces its market value. In 1999, both diseases were present in most crops at low levels and were not expected to cause major losses. However, grain harvested from many crops had a high proportion of discoloured seed. From visual inspection of the grain it was unclear whether the discolouration was caused by weather damage following rainfall just prior to harvest or by disease.

The studies reported here were undertaken to establish the cause of seed coat discolouration in faba beans and to determine the health of discoloured seed.

Material and Methodsw

Seed Source

A 100 kg sample was taken from a commercial seed lot of faba bean cv Fiord, harvested from a crop grown at Drung (near Horsham) in the Wimmera region of Victoria in 1999.

Seed from the sample was sorted into 6 categories according to Madeira *et al.* (3) based on the percentage of the surface area of each seed that was discoloured.

Seed Testing in the Laboratory

An agar plate method was used to determine the presence of fungi on faba bean seed. For each category of discoloured seed, 200 seeds were taken at random for testing. Seeds were surface sterilised (5 min in 0.5% NaOCl) before being placed in Petri dishes (5 per dish) containing potato dextrose agar (Merck 1.1030 39 g/L). Seeded dishes were then incubated for 14 days at 21 ± 2 °C in alternating 12 h cycles of near ultra-violet light (320-420 nm) and darkness. After incubation, seeds were examined and seed-borne fungi were identified.

The standard blotter test recommended by the International Seed Testing Association (2) was used to determine the percentage germination of each seed lot. For each category of discoloured seed, 100 seeds were taken at random, laid out on moistened absorbent paper, which was then wrapped up in a wet cotton towel and placed inside a plastic bag. The paper was kept moist for 21 days at 17.5 ± 2.5 °C. Following incubation for 21 days the numbers of germinated seeds were counted.

Glasshouse Experiment

A glasshouse experiment was conducted to determine the effect of seed discolouration on seedling emergence under controlled conditions. For each category of seed, 50 seeds were planted (5 seeds per pot) in 20 cm diameter plastic pots containing a composted pine bark potting mix. Pots were placed in a glasshouse with a temperature range of 15-25 °C and watered to field capacity daily. The numbers of fully emerged seedlings were counted at weekly intervals with a final count taken after 5 weeks.

Field Experiment

All categories of infected seed were sown at 100 kg/ha in 6-row plots, 7 m long with a row spacing of 15 cm, at Horsham (friable grey clay, pH 7.9 in $CaCl_2$) on 23 May 2000. Experimental design was a randomised block with 4 replicates. Emergence counts were made 3 weeks after sowing and plants were assessed for disease on 23 June and 25 August. Plants were assessed for the severity of ascochyta blight infection on leaves according to Bernier *et al.*

Discussion

Results of these studies show that discolouration of faba bean seed, in addition to reducing its market value, can also reduce seedling emergence and plant vigour which is expected to reduce grain yields.

Ascochyta blight appears to have been the main cause of seed discolouration in faba beans grown in 1999 despite the relatively dry conditions in many regions during spring. Seed infection is likely to have occurred late in the growing season, following heavy rains just prior to harvest. In the future, if there are heavy rains forecast late in the growing season, faba bean growers may be able to avoid seed discolouration by applying fungicides just prior to the rain. A fungicide trial conducted at Frances, South Australia, in 1999 demonstrated a significant reduction in seed staining due to ascochyta blight when an intensive spray regime was followed from the commencement of flowering (W. Hawthorne, pers. comm.). Further research is required to establish the optimum rates and timing of fungicide sprays required to prevent ascochyta blight staining and to determine the association between seed staining and seasonal conditions. In the past, faba bean producers have usually applied foliar fungicides to their crops early in the season to control ascochyta blight and have obtained little benefit from applications of foliar fungicides late in the season.

The high frequency of isolations of *A. fabae* from discoloured faba bean seed harvested in 1999, suggests that infection by ascochyta blight was the main cause of the discolouration. Where more than 25% of the seed coat was discoloured, germination, seedling emergence and plant vigour were significantly reduced, however, low levels of seed discolouration did not appear to have an adverse affect. Faba bean producers are advised not to use discoloured seed, particularly seed with over 25% discolouration, as it may seriously reduce the grain yield of their faba bean crops.

Germination

Germination is the process where growth emerges from a period of dormancy. The most common example of germination is the sprouting of a seedling from a seed of a flowering plant or gymnosperm. However, the growth of a hypha from a fungal spore is also germination. In a more general sense, germination can imply anything expanding into greater being from a small existence or germ.

Consumption of Seedlings

Seedlings are commonly eaten as a health food. These seedlings are usually labelled sprouts, but in a botanical sense are actually seedlings. There is controversy in whether seedlings, or sprouts, are really worth eating. These seedlings are many times used as organic foods eaten for their concentrations of certain vitamins the seedling naturally contains. This can be viewed out of proportion as the seedling is usually too small to contain enough vitamins or minerals to be in line with the amount they are claimed to obtain.

Stratification of Seeds

Seeds must be mature and environmental factors must be favorable before germination can take place. When a mature seed is placed under favorable conditions and fails to germinate, it is said to be dormant. The length of time plant seeds remain dormant can be reduced or eliminated by a simple seed treatment called stratification. Seeds should be planted promptly after stratification.

Seed Germination

The seed of a higher plant is a small package produced in a flower or cone containing an embryo and stored food reserves. Under favourable conditions, the seed begins to germinate, and the embryonic tissues resume growth, developing towards a seedling. The part of the plant that emerges from the seed first is termed a radicle. In some definitions, the appearance of the radicle marks the end of germination and the beginning of "establishment", a period that ends when the seedling has exhausted the food reserves stored in the seed. These are critical phases in the life of a plant. The mortality between dispersal of seeds and completion of establishment can be so high, that many species survive only by producing huge numbers of seeds. Some seeds require particular conditions to germinate, such as the heat of a fire (e.g., many Australian native plants), or soaking in a body of water for a long period of time. The germination index can be used as an indicator of phytotoxicity in soils. Some seeds will not germinate (begin to grow) until they have been dormant for awhile.

Requirements for Germination

Every variety of seed requires a different set of variables for successful germination. Typically these involve the proper levels of water, oxygen, temperature, and the correct soil conditions. Most seeds respond the best when water levels are enough to moisten the seeds but not soak them, when temperatures are slightly higher than room-temperature, and when oxygen is readily available. However, as stated earlier, this depends greatly on the individual seed variety. For example, some seeds will only germinate when temperatures reach in the hundreds of degrees (after a forest fire).

General Approaches to Promoting Seed Germination

Deciding appropriate germination test regimes for accessions so that germination tests reflect true viability with minimum interference from dormancy (and other factors which limit germination, such as hardseededness) can be one of the more difficult decisions which gene bank staff need to make. In the following chapters we have summarised useful information on the germination response of seeds of various genera to very many test regimes and dormancy-breaking treatments together with advice on suggested test procedures. In many cases, however, the information available is far from complete and accordingly much of the advice is very tentative. In such cases it is hoped that gene bank staff will be able to improve on the suggestions made. In order to help in this endeavour, in this chapter we discuss four different types of approach to determining appropriate germination environments.

Ecological Guidelines to Developing Appropriate Germination Environments

A consideration of the ecology of a species may sometimes help to provide some guidance as to appropriate germination test conditions. At the simplest level tropical species generally require higher temperatures for germination than temperate species. However, there are other more subtle responses which relate to the strategy of the plant in relation to its natural environment and a consideration of these may help in the development of germination test conditions which minimise dormancy.

Dormancy may prevent seeds from germinating in the wrong place at the wrong time

The majority of mature seeds show innate (or inherent or primary) dormancy — a condition necessary to prevent germination on the mother plant before the seed is shed. Innate seed dormancy in most species tends to be gradually lost with time (except in some tree species), often at a rate depending on temperature and moisture content. But induced (or secondary) dormancy, in many respects similar to innate dormancy, may subsequently be induced by conditions which signal that the current environment is inappropriate for germination — e.g. high temperatures or anaerobic conditions. Furthermore enforced dormancy may be temporarily imposed by conditions which again appear to play a role in ensuring that the seed only germinates when there is a reasonable probability of the seedling growing to maturity. Dormancy in all its forms therefore tends to prevent seeds from germinating in the wrong place at the wrong time.

Field Environment

For example, many annual or ephemeral species produce large numbers of small seeds. At any one time it is common for there to be more dormant seeds in the soil waiting for appropriate conditions than there are growing plants. Since small seeds only contain small food reserves, only a small amount of seedling growth can occur before the seedling needs to start photosynthesising. Thus germination more than a few centimetres below the soil surface or beneath dense vegetation could lead to seedling destruction. It would seem that many species have evolved dormancy mechanisms which avoid these problems. The seed needs to be able to respond to environmental factors that signal when it is at or near the soil surface. There seem to be two major environmental factors which many species use as indicators — light and alternating temperatures; and in most cases the seeds respond to both. Light is obviously only present at or near the soil surface; and diurnal temperature alternations have a maximum amplitude (or diurnal range) at the soil surface: the range rapidly diminishes with depth in the soil profile. Furthermore even at the soil surface the amplitude is diminished by the presence of vegetation. Since vegetation contains chlorophyll which absorbs strongly in the red region of the spectrum, the quality of light beneath vegetation may be inhibiting since far-red light penetrates much more and thus the light quality tends to affect the balance of phytochrome in the seeds in favour of the inactive Prform.

Accordingly it is very common for small seeds of annuals and ephemerals to be light-sensitive and to germinate in response to white (or red) light and alternating temperatures. In many cases neither of these factors on their own has a major effect and both are necessary for a maximum response. Both these factors also often interact with nitrate ions.

Maternal Environment

It may be worth considering the maternal environment in which the seeds develop as a further guide to distinguishing those species likely to require light treatments in order to promote seed germination. The light-filtering properties of the maternal tissues which surround the developing seeds, whilst they remain moist, can affect the sensitivity of seed germination to light in subsequent environments. Where the maternal tissues have a high chlorophyll content (that is are green) the seeds tend to require a light stimulus for germination (in subsequent tests). In contrast those seeds which have developed within maternal tissues where the chlorophyll content declined whilst the seeds remained moist tend to be able to germinate subsequently in light or darkness (that is are light insensitive). This is because the seeds developing within green maternal tissues would have received light filtered by the chlorophyll. This light would have a low red/far red ratio and most of the phytochrome would be in the inactive Pr form. In the latter case, however, the light would not have been filtered to such an extent; the red/far red ratio of the light received by the seeds would have been greater and thus more of the phytochrome would be in the active Pfr form which promotes germination. This Pfr would remain after desiccation and subsequent imbibition and accounts for the insensitivity of these seeds to light.

Large-seeded Species of Arid Environments

As mentioned in this study, sustained and/or high-intensity light can also inhibit the germination of some seeds. This response is more common amongst somewhat larger seeds of species of arid environments, where it is more important for seeds to germinate below the soil surface where short-term severe water shortage is less likely, and the probability of the shoot not reaching the soil surface before food reserves are reduced is lessened by the larger food reserves of the seed. In desert regions where rains are infrequent and of limited duration, ephemeral plants have short life cycles and it is important for them to germinate only after a substantial fall of rain (and not after a minor shower which would not provide sufficient water for survival). In such cases seeds of some species appear to have their own 'rain gauge' which depends on the fact that it takes a certain minimal amount of rain to leach germination inhibitors. Thus in such cases germination may be stimulated by washing the seeds in running water.

Probability of Plant Establishment and Survival

Dormancy mechanisms often also tend to ensure that the seeds do not germinate at times when the probability of the survival of the plant to maturity is poor. For example, many seeds in temperate latitudes tend to germinate after the winter, i.e. at the beginning of the main growing season. These seasonal considerations are often independent of seed size and thus many temperate species germinate best after a period of stratification (i.e. cool temperatures applied to moist seeds). Such a response ensures that seeds do not germinate at the beginning of winter. In contrast, seeds of species from Mediterranean climates tend to avoid germination at the beginning of the hot, dry summer and germination at the beginning of the cool moist winter is more common. In such cases it would be less likely to find marked responses to stratification. This would also be generally true of tropical plants, but see the note below.

In tropical rain forests the temperature and water supply is always adequate for growth and so seeds of such species often show little dormancy. Seeds of many of the woody perennial species are also recalcitrant. Such seeds do not have to survive long periods in the dry conditions, and fresh supplies of short-lived seeds are produced regularly by the mature plants. Many seeds of this type show little dormancy, and many are killed by cool temperatures (e.g. less than 10° to 15°C) which they would never normally experience. However, even in tropical forests some species are not recalcitrant, and some may show light responses to ensure that they only germinate when a gap arises in the forest canopy and/or the soil is disturbed after clearance, thus ensuring that early growth is not inhibited by low light intensity.

There are, of course, exceptions to these general principles. In particular it should be noted that laboratory treatments which emulate environments which the seeds would never experience in their natural habitat may sometimes promote germination. For example, low temperatures applied to moist seeds of some species of tropical and sub-tropical ecotypes can promote germination. Nevertheless the few examples above may be sufficient to indicate that a consideration of the ecology of the species may give useful clues as to the type of laboratory germination techniques which may be most appropriate.

The Anatomical Approach to Developing Appropriate Germination Environments

This approach has been developed largely by B.R. Atwater from long experience of attempting to germinate seeds of flower and ornamental species. Atwater's observations suggest that dormancy and germination requirements in these species are closely related to the maturity of the embryo and the structure and permeability of the seed coat coverings in the mature seed.

Atwater has described eight basic forms of seed structure. These categories have been described in this study. Within each of these categories it is suggested that the seeds have similar dormancy and germination patterns, but it should be emphasised that the mechanisms suggested are hypotheses rather than established facts.

Seeds with Eominant Endosperm (ENDOSPERMIC SEEDS) and Immature Dependent Embryos

Basal Rudimentary Embryos

The rudimentary embryos are the principal cause of delayed germination in this group. Inhibitors found in the endosperm may also contribute to the delay. Treatment consists of neutralization of the inhibitors with leaching, or with extra oxygen at low temperature followed by higher temperatures favourable to rapid embryo growth. Gibberellic acid increases the rate of embryo development if added to the substrate.

Axillary Linear Embryos

Linear embryos are similar to the above. An additional block is added by the seed coats which may limit oxygen entry, but seed coat permeability may be increased when the seeds are exposed to light. Gibberellic acid added to the substrate increases the rate of embryo development.

Axillary Miniature Embryos

Miniature seeds having minimal embryos are protected by thin seed coats showing a light requirement for oxygen permeability. Germination is prompt. Gibberellic acid is an aid, but not a substitute for light.

Peripheral Linear Embryos

Peripheral embryos which surround the food storage organ rather than being contained within it have multiple seed coverings which contain inhibitors, and which form barriers to both oxygen and water. Treatment is primarily leaching of the inhibitor but cold or light treatments may be required to increase permeability. Removal of the seed covering structures is also effective.

Seeds with only Residual or no Endosperm and Mature Independent Embryos

Hard Seed Coat

Hard (impermeable) seed coats are present which limit the entry of water for imbibition. Treatment for allowing water entry is necessary and may be mechanical scarification, acid treatment, percussion, high temperature treatment or soaking.

Thin Seed Coat with Mucilaginous Layer

Mucilaginous seed coats limit oxygen availability to the embryo after imbibition. Gibberellic acid is the most effective additive but light and potassium nitrate are also helpful, particularly when treatment with the two agents is combined.

Woody Seed Coat with Inner Semi-permeable Layer

Woody-textured and membranous multi-layered seed coverings cause blocks to germination which are most difficult to overcome. They readily admit water for imbibition but contain strong inhibitors which do not leach readily. Some of these are probably located in the thin residual endosperm surrounding the embryo or within the cotyledons of the embryo and are blocked from leaching by the semi-permeable membranous testa incorporated in the coverings. Excised embryos germinate promptly.

Fibrous Seed Coat with Separate Semi-permeable Membranous Coat

The Compositae family forms a specialised group which is similar to the preceding section. Embryos contain inhibitors in their cotyledons. Testa and thin endosperm form a separate membranous semi-permeable coat. An outer fibrous coat offers various dispersal forms. Extracted embryos grow promptly if leaching is complete.

Atwater has developed her ideas and experience almost to the stage of providing species prescriptions. Much of this information is summarised in subsequent chapters, but the problems inherent in species prescriptions are discussed below. Note also that certain plant families have species in more than one category of seed anatomy.

Prescriptions for Germinating All Accessions of a Given Species

This approach to determining germination test environments is exemplified by the Rules of the International Seed Testing Association (ISTA) and the Association of Seed Analysts of North America (AOSA). Commercial seed testing has evolved over the past 130 years to provide a certain degree of assurance to farmers, growers and foresters that any seeds of agricultural, horticultural or arboreal crops that they may purchase are capable of germinating under normal field conditions, emerging above the soil surface and developing into plants which yield an adequate crop. Thus, by assessing the quality of seed lots before they are sown, commercial seed testing minimises the risk to the farmer, grower or forester of sowing seed lots which do not have the ability to produce an abundant crop. Since this assessment will be done by only one of many laboratories and since there is also considerable international trade in seed of many species, assessments by different seed testing laboratories must give similar results.

Compatibility Between Seed Testing Stations

To achieve compatibility of germination test results between different laboratories ISTA and AOSA co-ordinate and produce agreements between seed testing laboratories on the procedures to be applied for each species. We have discussed in this study how the proportion of seeds which will germinate in a germination test can be affected by the test environment. Thus if seed testing laboratories used very different germination test conditions the results obtained in different laboratories could differ substantially. To aid compatibility ISTA and AOSA seek agreement between laboratories as to the most appropriate environment for testing the germination of a given species. In theory this implies agreement upon a single germination test environment. In practice alternative environments are often specified for a given species. This is either because of disagreement between laboratories as to the single most appropriate environment, or because of the need to provide alternative procedures for certain seed lots, or because of practical difficulties in providing certain environments (for example, alternating temperatures) in some laboratories.

The rules of the ISTA and AOSA list the prescribed germination test conditions for each species. These prescriptions specify the substrata, temperatures and test duration which seed testing laboratories must adhere to, Additional directions are provided for special treatments which may be required for certain species together with recommendations for breaking dormancy. Now, these rules and directions are intended to give the most regular, rapid and complete germination for the majority of samples of seed of a particular species. To that end we presume the rules are successful, but how closely does the above aim coincide with the problems facing those testing seeds within gene banks?

Prescriptions are not Available for All Species

First, ISTA and AOSA tend to only provide rules for those species in which certian individual seed testing stations have knowledge of the germination behaviour of a large number of seed samples. Thus the species listed in the ISTA and AOSA rules are those in which there is considerable trade in the sees; the seed-propagated crops. In contrast gene banks will often need to germinate seeds of species which are not yet in substantial trade in the developed world (e.g. the tropical pasture species), true seeds of vegetatively propagated crops (e.g. potato, Solanum tuberosum), and seeds of species which do not have crop status such as wild and weedy relatives of modern crop species (e.g. teosinte, Euchlaena mexicana). Thus the first drawback for gene banks of the ISTA/AOSA rules is that for many species on which gene banks require advice on germination test conditions no information is provided.

Prescriptions Developed for Modern Cultivars

Secondly, within those species for which ISTA and AOSA prescribe test conditions the range of seed lots which seed testing stations test for germination, and upon the results of which ISTA precriptions are based, are generally limited to relatively non-dormant, high quality, modern, genetically homogeneous cultivars. That is in most cases the cultivars with which the rules have been developed are likely to have been subjected to considerable selection pressures. In contrast gene banks will be testing seeds of primitive cultivars and genetically heterogeneous landraces which may also show considerable dormancy and/or be of poor quality. Thus the second drawback for gene banks of the ISTA/AOSA rules is that the prescribed conditions for a particular species will not necessarily be suitable for those acessions within gene banks which are dissimilar to modern cultivars — or of poor quality.

Relative Priorities of the Aims of Seed Testing

Thirdly, for some crops seed testing stations may operate under considerable pressure at harvest time. For example, in temperate countries autumn sown cereal seeds are sown only five or six weeks after harvest. Thus seed testers may need to complete germination tests in as short a time as possible. Hence seed testing stations' aim of providing conditions which give rapid germination. Now in at lest one group of species — the temperate cereals — although the conditions prescribed by the rules may give rapid germination this is only achieved in our experience by using supra-optimal temperatures at the cost of the parallel aims of achieving regular and complete germination which are better achieved at much lower temperatures. In contrast to seed testing stations, gene banks are not under any pressure to give priority to rapid germination over achieving regular and complete germination. Thus the third drawback for gene banks of the ISTA/AOSA rules is that the aims of rapid and complete germination may not always be compatible. Note that this criticism does not apply to the rules for all crops. In the case of tree species, for example, considerable periods are allowed for dormancy-breaking and germination test treatments by ISTA and AOSA rules, and the requirement for complete germination takes precedence over the requirement for rapid germination.

Field Planting Value

The purpose of commercial testing is to provide information on the comparative field planting value of different seed lots. In general the ISTA and AOSA prescriptions make no attempt to enable those viable seeds which will not germinate under field conditions to germinate in the laboratory test. If, for example, dormancy is a problem in laboratory tests, but not in field sowings (for example, due to differences in temperature between the two regimes) then directions to break dormancy must be provided in the rules — otherwise the field planting value may be underestimated. However, where dormancy is a problem in both laboratory tests and field sowings then to obtain an indication of field planting value the rules may not necessarily provide dormancy-breaking directions — since the removal of dormancy in laboratory tests may result in an overestimate of field planting value.

Again whether or not this objective of ISTA or AOSA prescriptions and recommendations makes them unsuitable for viability testing in gene banks depends on the species. For example, in tree species directions for dormancy removal are suggested to enable viability to be estimated in laboratory tests and these directions must also be followed in field sowings if laboratory results are to provide a reasonable estimate of field planting value. In contrast in some species where hardseededness prevents germination no recommendation is made to render the seed coat permeable. Instead the proportion of hard seeds is reported. Thus the fourth drawback of ISTA/AOSA rules is that no attempt may be made to overcome a wide range of physiological phenomena which may prevent the germination of viable seeds under both laboratory and field conditions. Of these phenomena the most important are hardseededness and strong dormancy. In contrast to seed testing stations, gene banks aim to provide conditions which will enable all viable seeds to germinate both in laboratory tests and in subsequent field or glasshouse sowings.

Other Sources of Prescriptions

Numerous groups of workers have attempted to provide prescriptions for germination tests of species for which no ISTA/AOSA rules are available or which overcome certain of the deficiencies of the ISTA and AOSA prescriptions for species where rules are available. Particular emphasis has been placed on overcoming dormancy since this can be an important problem for plant breeders, and also for seed testers if considerable dormancy is experienced when the seeds are tested but which is likely to be lost during storage before sowing.

However, much of this information is also of limited use in gene banks. This is because the majority of such information relates to the application of single dormancy breaking agents alone. It was explained that if this approach is used then the treatment concentration (if a chemical agent) or duration (if another form of treatment) required to break dormancy in the most dormant seeds would be likely to damage less dormant seeds, particularly if these are also of poor quality.

The Aim of Developing a Species Prescription for Use in Gene Banks

We believe that the ultimate desirable aim of a prescriptive test for a given species is to be able to provide within one test environment or procedure those conditions which enable all viable seeds of that species — whether strongly dormant, moderately dormant, weakly dormant or non-dormant, and of high or low quality — to be able to germinate. This means that any stimulus or combination of stimuli used to promote the germination of dormant seeds must not be damaging to other non-dormant seeds in the sample or to intolerant genotypes.

Developing a Prescription for a Species

Whether this aim can in fact always be achieved is a matter for debate, but two considerations are important in attempting to achieve it. The first is to ensure that the germination test temperature is suitable for poor quality seeds; within this range attempts can then be made to determine the most suitable temperature regime which will minimise the expression of dormancy. In certain species this first step may result in an adequate single germination test environment without any need for further treatment. In temperate cereals this appears to be the case, where a single low constant temperature is appropriate for both purposes. However, in other species simple solutions of this type may not be available and so it is necessary to take a second step and provide further stimuli to break dormancy which do not impair the germination of less dormant seeds. In our opinion this is often best attempted by combining more than one additional stimulus with all stimuli being applied at relatively low dose-levels. This is an attempt to utilise the positive interactions observed between many dormancy-breaking agents. The treatments developed for Oryza spp. and Vitis spp. provide examples of this type of procedure.

Gene banks considering developing their own germination test prescriptions may be discouraged by the wide range of different treatments — and the even wider range of different treatment combinations — which may be of potential benefit in promoting the germination of viable seeds. Consequently it is worthwhile considering which factors should receive priority in any investigation. We believe that the three most important factors to consider first are to ensure that the seeds have imbibed moisture without damage and then to attempt to determine the most suitable temperature and light regimes for germination.

Provided a prescription is already available for a species which is suitable for application in a gene bank then this would be the simplest approach for gene banks to use. However, even where a prescription is well-tried, gene banks will still need to be alert to the possibility that some accessions may be discovered for which the prescription does not appear to work. Furthermore, for many years to come, there will continue to be a large number of species for which there are no adequate prescriptions. Accordingly alternative approaches to providing suitable germination test environments are also considered in this handbook.

Taxonomic Algorithms to Determine Appropriate Germination Test Environments

As mentioned above, germination test prescriptions are available for only a small proportion of seed-producing species. What environment should be chosen to test the germination of seed accessions of species for which no prescription or no acceptable prescription is available? This is exactly the problem which faced staff at the Royal Botanic Gardens Kew, Wakehurst Place Gene Bank, for almost all their accessions, because this gene bank deals entirely with seed collections of species which do not have great commercial significance. The method that they developed to answer the question posed above provides another approach to developing appropriate germination test regimes for seed accessions maintained in gene banks.

Algorithm: the Shortest Series of Tests to Define a Suitable Test Environment

This approach acknowledges that the most suitable environment for testing each accession is unknown. Thus, instead of a single prescription, an attempt has been made to develop the shortest series of tests which is likely to give a solution to the problem of defining a suitable test environment. The decision as to which test to try next depends on the results obtained in the previous test(s), and so we call this series of tests an algorithm. For gene banks a truly universal algorithm would be one appropriate to all seed-producing species, irrespective of taxonomy. However, such an approach would be rather unwieldy and require a large number of tests. Certain similarities in the response of seed germination to environment for species within a family can be recognised. Consequently this approach provides a separate algorithm for each family where sufficient knowledge has been accumulated. Where available, the algorithms developed by the staff at the Wakehurst Place Gene Bank are presented according to family. However, it is necessary here to provide some explanation of how the algorithms were developed, their possible faults, and how they should be applied.

Use of Tetrazolium Test to Determine Efficacy of a Prescription

Although a gene bank can never be absolutely certain about the efficacy of any approach to germinating seeds, do not forget that the efficacy of a species prescription on a particular accession can be tested by comparing the results of the germination test with that of the topographical tetrazolium test on a sub-sample of the seed lot and by also applying a tetrazolium test to the seeds remaining ungerminated at the end of the test. In this way one can determine the proportion (if any) of viable seeds which are either killed by the germination test procedure or remain dormant at the end of the germination test procedure. In the above we have detailed objections to the wholesale and unquestioning adoption of either ISTA or AOSA prescribed germination test procedures as gene bank germination tests. However, this certainly does not mean that such prescriptions should be ignored by gene banks. The ISTA and AOSA rules summarise a wealth and depth of expertise and experience which is unmatched. Thus where available, the prescribed conditions represent a control against which any proposed alternative should be tested; and, despite our earlier comments, certain of these prescriptions will be suitable for gene bank use.

The Development of the RBG Kew Wakehurst Place Algorithms

Storage at the Wakehurst Place Gene Bank began in 1974. Since then all accessions entered into storage have been tested for germination as soon as possible thereafter. In many cases accessions were tested in more than one germination test environment, the reason for this being an inadequate proportion (less than 85%) of seeds germinating in the first test. Thus for many accessions it was possible to compare germination test results over a number of different environments with and without the imposition of various additional dormancy-breaking treatments. For each accession the best germination test regime was noted, and this regime was then used for all subsequent germination tests of that accession. In addition the results of all the different germination tests carried out on each accession were filed according to species to provide a growing information bank on the response of seed germination to different test conditions.

At 21°C all seeds exhibited dormancy. The alternating temperature 'regime of 21°/11°C was only partly successful in breaking this dormancy. Therefore towards the end of the latter test, a decision was made to sample more seeds from the accession and begin two additional germination tests in different conditions (which are known to promote germination in several species). One of these additional regimes — consisting of a preliminary period during which the imbibed seeds were exposed to 26°C (sometimes described as a warm stratification treatment) followed by subsequent exposure to 11°C — was entirely successful at breaking dormancy, enabling all the seeds tested to germinate. Consequently this regime will be used for any subsequent germination tests to monitor the viability of this accession.

Over the past ten years or so a considerable amount of information on the response of seed germination to various test regimes (similar to that described above for Veronica verna) has been generated by staff at Wakehurst Place. This raw information has subsequently been analysed, principally by Mr. S. Linington, in the following manner. For a particular family all the routine test data for all accessions (each similar in form to that provided in were examined. It was (arbitrarily) decided that test regimes in which 85%, or more, of the full seeds germinated would be classified as successful. For each accession all the successful regime(s) (if there were any) were listed.

When this first stage of analysis was completed a pattern of successful regimes within each family was sought. Emphasis was placed on determining those simpler regimes (that is those employing few dormancy-breaking agents) which were successful for the greatest number of accessions. These regimes are those which are now used as the first step in the algorithms which have been developed. More complicated regimes (involving the use of many dormancy-breaking treatments) which are successful for fewer accessions are then provided as subsequent alternatives (later steps) in the algorithm if the first step does not provide a suitable environment for germination testing.

Possible Limitations of the Algorithms

There are a number of possible faults inherent in this method of analysis. First, by excluding from the analysis all accessions which failed to reach 85% germination under any test regime, the response of accessions which are either very dormant, or very poor quality and vulnerable in certain germination test environments may have been ignored. Secondly, it is based on tests of accessions within the Wakehurst bank which, at least in the past, were biased towards collecting in temperate regions. Not only may this result in an algorithm more suited to temperate collections, but it may also result in algorithms more suited to certain tribes within each family. For example, the algorithm for the Leguminosae was based largely on accessions of the Papilionoideae tribe and little data was available for accessions from either the Caesalpinoideae or Mimosoideae tribes. Finally it should be noted that the analysis was based on data obtained over only a limited range of treatments. Other, untested, treatments may be superior and there is as yet no way of knowing whether this will be the case.

Success of the Algorithms may Vary between Families

Despite these reservations, the algorithm approach to determining germination test environment works well at Wakehurst and is likely to be useful in other gene banks. Tests at Wakehurst on previously untested accessions using the algorithms for the Amaranthaceae, Gramineae, Leguminosae and Solanaceae families have shown that 100%, 50%, 90% and 75% of accessions within each family reached 85% germination or greater. In view of the importance of the Gramineae in food production the limited success of the algorithm for this family is disappointing. One group of species within the Gramineae whose seed is particularly difficult to germinate is the tropical pasture species. Unfortunately tests on five seed lots of tropical pasture species at Reading confirm that the present algorithm for the Gramineae is not particularly helpful for determining germination test environment in these species.

Demonstration of the Use of an Algorithm

Nevertheless, can be used to explain the method of applying the algorithm should be studied together with the Gramineae algorithm provided in this study. In every feasible combination of dormancy-breaking agents within the algorithm was applied to the seeds, amounting to some 24 separate test environments. Note, however, that only a few of these tests would have been applied if the algorithm had been followed. The pathway through the tests which would have been followed if the Gramineae algorithm had been applied is shown by the boxed values in, the sequence being from the top of the table to the bottom.

The Gramineae algorithm provided by the Wakehurst Place Gene Bank. First the germination test temperatures are not identical, but are within 1°C — except for the upper limit of the alternating temperature regime where the difference is 2°C. These minor differences are unimportant and result from differences in the standard temperatures available at Reading and Wakehurst. Secondly the comment in the algorithm to test at constant temperatures either below 20°C or above 25°C depending upon which gave the higher test result was not followed, since previous experience with dormant seed lots of these species indicated that alternating temperatures were essential to promote germination. Thirdly, following on from the last point, seeds were also tested for germination at the two additional alternating temperature regimes of 15°/30°C and 10°/30°C.

To enable the reader to understand fully the use of the algorithm, an explanation is provided below of the sequence of tests and decisions that would have been applied for tests on the accession of Brachiaria decumbens. The first stage in the algorithm was to test the seeds at constant temperatures of 20°C and 25°C. These test results were inadequate (less than 85% germination).

Now, according to the algorithm the next step would have been to look for trends in the results of these two tests (i.e. was germination greater at the higher or lower temperature?) and test for germination at either a more extreme higher or a more extreme lower constant temperature accordingly. This, however, was not done for the reasons given above and we proceeded to the second step of the algorithm which was to test a fresh sample of the seeds in an alternating temperature regime of 35°/20°C (the nearest regime available to the standard 33°/19°C Wakehurst regime). The germination obtained in this test was substantially greater than at either of the two constant temperatures, but again inadequate as a germination test regime. Consequently further tests were required.

The third step according to the algorithm was to test a fresh sample of the seeds in a medium containing 10^{-3} M KNO_3 at, in view of the difference observed between the results at steps 1 and 2, an alternating temperature of 35°/20°C. Again the result of this test at step 3 was an improvement on the result at step 2 but inadequate as a germination test regime since 85% germination was not achieved.

Thus the fourth step was to remove the seed covering structures from a fresh sample of the seeds and test in a medium containing 10^{-3} M KNO_3 (since the result at step 3 was greater than the result at step 2) at an alternating temperature of 35°/20°C. Once again the result of the test at this step was an improvement on the previous test result (suggesting that dehusking was a worthwhile treatment), but the percentage of seeds germinating was still unsatisfactory.

Consequently the final step in the algorithm was applied. Namely dehusked imbibed seeds were pre-chilled at 3°-5°C for 8 weeks and then transferred to an alternating temperature regime of 35°/20°C in a medium containing 10^{-3} M KNO_3. Unfortunately this test resulted in substantially fewer seeds germinating than the previous test. Thus the conclusion reached by using the algorithm is that dehusked seeds of this accession of Brachiaria decumbens should be tested on a substrate containing 10^{-3} M KNO_3 in an alternating temperature regime of 35°/20°C.

Note that the decision making procedure at each step in the algorithm did not result in the algorithm missing more favourable treatment combinations. For example, if the algorithm had been followed dehusked seeds would not have been tested at 35°/20°C without 10^{-3} M KNO_3 in the germination medium. In fact this treatment gave a very similar result (that is there was no significant difference) to the treatment with dehusked seeds which did include 10^{-3} M KNO_3 in the germination medium. Thus it appears that the use of 10^{-3} M KNO_3 in the medium is in fact not necessary, but its use (apparently necessary if the algorithm had been followed) does not result in a significantly lower proportion of seeds germinating. In that sense we can regard the decision-making procedure within the algorithm as being successful, but unfortunately the algorithm itself was not successful since the level of viability indicated by the tetrazolium test was not achieved in these germination tests.

Sensitivity of Germination to the Temperature Regime

In addition to all the possible germination test regimes according to the Gramineae algorithm, also presents results of tests carried out in two additional alternating temperature environments (15°/30°C and 10°/30°C) and also in combination with 10^{-3} M KNO_3 (where sufficient seeds were available). These results suggest that, for some seed lots at least, those gene banks testing tropical grass species would be well advised to include additional alternating temperature regimes as the second step of the algorithm. For three of the seed lots tested here small differences in the alternating temperature regime greatly influenced the proportion of seeds which germinated. Not only did this result in almost full germination of viable seeds, but it would have avoided the need for several stages of the algorithm in the case of Gatton Panic.

Possible Use of Ungerminated Seeds from Prior Step in Algorithm for the Current Test

In the algorithms provided here additional tests are carried out by drawing a fresh sample of seeds from the accession bulk in storage. In some cases, for example if the supply of seeds is severely limited, it is possible to test the response to additional treatments by applying these to those ungerminated seeds which remain fresh and firm at the end of a previous test. This is particularly suitable for scarification, dehusking or puncturing treatments since it may avoid the need to apply these time-consuming procedures to all seeds tested. For example, if 70% of seeds tested of a legume accession germinate without scarification it is easier to scarify the remaining proportion of seeds and to return these to the germination environment than to draw a fresh sample of seeds from storage and scarify all of the seeds to be tested.

Duration of Germination Tests

During each step of the algorithm it is necessary to decide when to terminate the germination test and whether to pursue the next step in the algorithm. Since both aged and dormant seeds may only begin to germinate in a test after a substantial delay it is not possible to provide categorical advice. This technique can be used for each step in the algorithm, but each test should not be continued for an excessive period because the next step in the algorithm might result in a more suitable and more rapid procedure. That is if the progress of germination is very slow it might be better to pursue the next step in the algorithm.

A minimum germination test period of 28 days is suggested. A comparison of results after 28 and 56 days in test, for example, shows substantial germination after 28 days in test for only the Brachiaria humidicola accession tested at 35°/20°C. Consequently it is suggested that the decision to test a further sample of seeds in the next step of the algorithm can generally be made once the 28-day test result is known, but that the current test be continued for a further period dependent upon the shape of the germination progress curve. In tests at a very low temperature (e.g. 6°C), however, it may be better to delay a decision as to whether to continue through the algorithm until the 56-day test result is known.

Use of Algorithms at Accession Receipt

It is suggested that the algorithms could be applied soon after an accession is received if the standard advice on germination test procedures and dormancy-breaking treatments is unsatisfactory or if no advice is available. The number of seeds tested in each of the regimes required by the algorithm need not exceed 100 (and 50 seeds may well be sufficient), but the seeds should be sampled from the accession at random and replication in each test is advisable.

Once a suitable test procedure has been developed a further sample of 200 seeds can be withdrawn from the accession and tested by this procedure in order to 'estimate initial accession viability. Subsequent monitoring tests of accession viability will also apply the same germination test procedure, but may be either fixed sample size tests or sequential tests. Note that it is envisaged that the algorithms would only be applied once, i.e. when the accession is first received: the subsequent monitoring tests would employ the most appropriate regime already determined.

Reproduction in Plants

Reproductive Strategies

A. A life cycle is full sequence from fertilization and formation of a zygote to gamete formation once again.

1. In contrast to animals with one type of adult generation, flowering plants exhibit an alternation of generations life cycle that includes a diploid and a haploid generation.

2. Sporophyte is diploid generation in an alternation of generations life cycle.

   a. Sporophyte produces haploid spores by meiotic division.

   b. Spores develop into a haploid gametophyte.

3. Gametophyte is haploid generation in an alternation of generations life cycle.

   a. Gametophyte produces haploid gametes by mitotic division; gametes fuse to form diploid zygote.

   b. Zygote undergoes mitotic cell division to develop into the sporophyte.

4. In flowering plants, diploid sporophyte is dominant (longer lasting); it is what we commonly recognize.

5. Sporophyte is generation that contains vascular tissue and has other adaptations suitable to living on land, including production of flowers.

6. A flower produces two types of spores, microspores and megaspores.

   a. A microspore is a plant spore that develops into a microgametophyte.

      1) Microgametophyte is pollen grain; wind or animals carry it to megagametophyte.

      2) When mature, its non-flagellated sperm cells travel down pollen tube to megagametophyte.

   b. Megaspore is a plant spore that develops into a megagametophyte, the embryo sac which remains within a sporophyte plant.

7. Alternation of generations is modified so sperm does not require water to reach an egg, which nonflowering plants may require.

a. Gametophytes are microscopic and dependent upon the sporophyte.

b. Microgametophyte is transported to megagametophyte by wind or an animal; water is not needed.

c. Pollen tube carries a sperm to an egg; again no outside water is needed.

d. Following fertilization, diploid zygote develops into an embryo within a seed enclosed by a fruit.

B. The Sporophyte

1. A flower is the reproductive organ of a flowering plant; it develops within a bud.

2. Shoot apical meristem stops forming leaves to form flowers; axillary buds can become flowers directly.

3. Flower structures are modified leaves attached to a stem tip (receptacle).

4. Monocot flower parts are in threes or multiples; dicot flower parts are in fours or fives or multiples.

5. Sepals are leaf-like, usually green; this outermost whorl protects the bud as flower develops within.

6. Petals are interior to sepals; coloration accounts for attractiveness of many flowers.

a. Size, shape, and color of a flower are attractive to a specific pollinator.

b. Wind-pollinated flowers often have no petals at all.

7. Pistil is vase-like structure located at center of a flower; it contains one or more carpels.

a. Pistil may be simple or compound.

b. Simple pistil contains one carpel; compound pistil contains multiple carpels which are often fused.

8. Carpelsare reproductive units of flowers and have three parts.

a. Stigma is an enlarged sticky knob on end of a style; stigma serves to receive pollen grains.

b. Style is a slender stalk that connects stigma with the ovary.

c. Ovary is enlarged base of a carpel that contains a number of ovules.

9. Grouped about a pistil are stamens, stalked structures that have two parts.

   a. Anther is a sac-like container within which pollen grains develop.

   b. Filament is a slender stalk that supports the anther.

10. Not all flowers have sepals, petals, stamens, and a pistil.

   a. Complete flowers have sepals, petals, stamens, and a pistil; incomplete flowers do not.

   b. Perfect flowers have both stamens and a pistil.

   c. Staminate flowers have only stamens.

   d. Pistillate flowers have only pistils.

11. If staminate and pistillate flowers are on same plant, the plant is monoecious.

12. If staminate and pistillate flowers are on different plants, the plant is dioecious.

13. It is not strictly correct to call a pistil the female part of a flower, or a stamen the male part; these organs do not produce gametes.

14. The pistil and stamen produce megaspores and microspores that mature to produce eggs and sperm.

C. The Gametophytes

1. The ovary in a carpel contains one or more ovules.

2. Ovules has center mass of parenchyma cells covered by integument with one opening (micropyle).

3. One parenchyma cell enlarges to become megasporocyte.

   a. Megasporocyte undergoes meiotic cell division to produce four haploid megaspores.

   b. Three megaspores disintegrate; one megaspore nucleus divides mitotically into eight nuclei.

   c. When cell walls form around the nuclei later, there are seven cells, one of which is binucleate.

   d. Megagametophyte (embryo sac) consists of seven cells:

      1) one egg cell,

      2) two synergid cells,

      3) one central cell with two polar nuclei, and

      4) three antipodal cells.

4. An anther has four pollen sacs; each contains many microsporocytes.

a. Microsporocytes are microspore mother cells.

b. Microsporocytes undergo meiotic cell division to produce four haploid microspores.

c. Each microspore divides mitotically forming two cells: a tube cell and a generative cell.

d. This immature microgametophyte is a pollen grain.

e. Eventually each generative cell divides mitotically to form two sperm.

f. Walls separating pollen sacs break down and pollen grains are released.

5. Pollen grains are windblown or carried by various kinds of animals to stigma of a pistil.

D. Pollinatión and Fertilization

1. Pollination and fertilization are separate events.

2. Pollination is strictly transfer of pollen from anther to stigma of a pistil.

a. Pollination occurs by wind or with assistance of particular animal pollinators

b. Self-pollination is transfer of pollen from anther to stigma of the same plant.

c. Cross pollination is transfer of pollen from anther of one plant to stigma of another plant; it is evolutionarily advantageous because of genetic recombination resulting in new and varied plants.

3. Fertilization

a. Fertilization is fusion of nuclei, as when sperm nucleus and egg nucleus fuse.

b. When a pollen grain lands on a stigma, it germinates, forming a pollen tube.

c. A germinated pollen grain, containing a tube cell and two sperm, is mature microgametophyte.

d. As a pollen tube grows, it passes between cells of stigma and style to reach micropyle of an ovule.

e. Pollen tube grows through micropyle and releases both sperm cells into the ovule.

f. One sperm nucleus unites with the egg nucleus, forming a 2n zygote.

g. The other sperm nucleus migrates and unites with polar nuclei of central cell, forming 3n endosperm nucleus.

h. This is double fertilization.

4. Zygote divides mitotically to become the embryo; endosperm cell divides mitotically to become endosperm.

a. Embryo, in most plants, is a young sporophyte.

b. Endosperm is tissue that will nourish embryo and seedling as they undergo development.

Seed Development

A. Stages in Development of Dicot Embryo

1. After double fertilization, the ovule develops into a seed.

2. The endosperm nucleus divides to produce a mass of endosperm surrounding embryo.

3. The single-celled zygote also divides, forming two parts: embryo and suspensor, which anchors embryo and transfers nutrients to it from the sporophyte plant.

B. Embryonic Development

1. After differentiation into embryo and suspensor, one or two cotyledons develop.

a. Cotyledon (seed leaf) provides nutrient for a developing plant before it photosynthesizes.

b. Dicot embryos develop two cotyledons; monocot embryos develop only one cotyledon.

c. During development of a monocot embryo, the cotyledon rarely stores food; rather, it absorbs food molecules from endosperm and passes them to embryo.

d. During development of a dicot embryo, cotyledons usually store nutrients the embryo uses, obtaining those nutrients from endosperm.

2. Embryo continues to differentiate into three parts.

a. Epicotyl is between cotyledons and first leaves; it contributes to shoot development.

b. Hypocotyl is below cotyledon and contributes to stem development.

c. Radicle is below hypocotyl and contributes to root development.

Fruit Types and Seed Dispersal

A. Seeds and Fruits

1. As a zygote develops into embryo, integuments of the ovule harden and become seed coat.

2. Ovule matures into the seed, containing sporophyte embryo plus stored food.

3. Ovary (and sometimes other floral parts) develops into fruit (mature ovary that usually contains seeds).

B. Types of Fruits

1. As fruit develops from an ovary, ovary wall thickens to become pericarp.

2. Simple fruit develops from an individual ovary, either simple or compound.

3. Fleshy fruit has a fleshy pericarp (e.g., peach, plum, olive, grape, tomato, apple, and pear).

4. Compound fruit develops from a group of individual ovaries (e.g. apple, tomato).

5. Dry fruit pericarp is dry (e.g., milkweed, pea, bean, lentil, poppy, sunflower, acorn, rice, and barley).

   a. Aggregate fruit develops from ovaries from single flower (e.g., blackberry), while an aggregate fruit where each ovary becomes a one-seeded fruit is called an achene (e.g., strawberry).

   b. A multiple fruit develops from ovaries from separate flowers fused together (e.g., pineapple).

C. Seed Dispersal

1. For plants to be widely distributed, seeds have to be dispersed away from parent plant.

2. Hooks and spines of clover, bur, and cocklebur attach to fur of animals.

3. Birds and mammals eat fruits, including seeds, and defecate them at a distance.

4. Squirrels and other animals gather seeds and fruits and bury them some distance away.

5. A coconut floats hundreds of kilometers; some plant seeds have trapped air or inflated sacs.

6. Woolly hairs, plumes, and wings disperse by wind.

7. Touch-me-not has seed pods that swell as they mature and burst, hurling their ripe seeds.

D. Seed Germination

1. Some seeds do not germinate until they have been dormant for a period of time.

a. Seed dormancy is a time during which no growth occurs even though conditions are favorable.

b. In temperate zones, seeds may have to be exposed to cold weather before dormancy is broken.

c. In deserts, germination requires adequate moisture; this ensures that seeds do not germinate until a favorable growing season has arrived.

2. Germination takes place if there is sufficient water, warmth, and oxygen to sustain growth.

a. Regulation of germination involves both growth inhibitors and growth simulators.

1) Fleshy fruits contain inhibitors; germination does not occur until seeds are removed and washed.

2) Growth stimulators are present in seeds of some temperate zone woody plants.

b. Mechanical action may also be required to facilitate germination.

1) Water, bacterial action, and fire act on seed coat, allowing it to become permeable to water.

2) Water uptake causes seed coat to burst.

3. Germination in Dicots

a. Prior to germination, embryo consists of the following:

1) two cotyledons that supply nutrients to embryo and seedling soon shrivel and disappear;

2) a plumule-a rudimentary plant consists of an epicotyl bearing young leaves;

3) hypocotyl, which becomes the stem; and

4) radicle, which develops into roots.

b. As dicot seedling emerges, the shoot is hook-shaped to protect delicate plumule.

c. As seed germinates in darkness, it etiolates; stem increases in length and leaves remain small.

d. Phytochrome pigment sensitive to red and far-red light induces normal growth in light.

4. Germination in Monocots

a. Endosperm is food-storage tissue; cotyledon does not have a storage role.

b. Monocot "seed" is actually the fruit; outer covering is the pericarp.

c. Prior to germination, embryo consists of one cotyledon, a plumule, and a radicle.

d. Plumule and radicle are enclosed in protective sheaths: coleoptile and coleorhiza, respectively.

e. Plumule and radicle burst through these coverings when germination occurs.

Asexual Reproduction in Plants

A. Means of Asexual Propagation

1. Plants contain non-differentiated meristem tissue and reproduce asexually by vegetative propagation.

2. In asexual reproduction, offspring arise from a single parent and inherit genome of that parent only.

3. Vegetative propagation utilizes meristematic tissue of a parent plant.

a. Nodes of stolons will produce strawberry plants.

b. Violet plants grow from nodes of rhizomes.

c. Each eye of a potato plant tuber is a bud that produces a new plant.

d. Sweet potatoes can be propagated from modified roots.

e. Many trees can be started from small "suckers."

4. Stem cuttings have long been used to propagate a wide array of plants (e.g. sugarcane, pineapple).

5. Discovery that auxin will cause roots to develop has expanded ability to use stem cuttings.

B. Tissue Culture of Plants

1. In 1902, German botanist Gottleib Haberlandt suggested producing entire plants from tissues.

2. Tissue culture is process of growing tissue artificially in a liquid culture medium.

3. Haberlandt stated plant cells were totipotent; each cell has full genetic potential of the organism.

4. In 1958, Cornell botanist F.C. Steward grew a complete carrot plant from a tiny piece of phloem.

5. When cultured cells are provided with sugars, minerals, vitamins, and cytokinin, the undifferentiated

cells divide and initially form a callus, an aggregation of undifferentiated cells.

6. The callus then differentiates into shoot and roots and develops into complete plants.

7. Micropropagation is a commercial method of producing thousands to millions of identical seedlings, by tissue culture in limited space.

8. Meristem culture micropropagates many new shoots from a single shoot apex culture in a medium with correct proportions of auxin and cytokinin.

a. Since shoots are genetically identical, adult plants that develop are clonal plants.
b. Clonal plants have same genome and display same traits.
c. Meristem culture generates meristem that is virus-free; plants produced are also virus-free.

9. Entire plants can be grown from single plant cells.

a. Enzymes can digest cell walls and produce naked plant cells called protoplasts.
b. Protoplasts regenerate a cell wall and begin cell division.
c. Clumps of cells can be manipulated to form somatic embryos.
d. Somatic embryos encapsulated in a hydrated gel ("artificial seeds") can be shipped anywhere.
e. Somatic embryos are cultured by the millions in large tanks (bioreactors).
f. Plants generated from somatic embryos vary because of mutations; these somaclonal variations may produce new traits.

10. Anther culture cultures mature anthers in a medium of vitamins and growth regulators.

a. Haploid tube cells within a pollen grain divide, producing proembryos made of up to 40 cells.
b. Finally, pollen grains rupture, releasing haploid embryos.
   1) The researcher can then generate a haploid plant.

2) Chemical agents are added to encourage chromosomal doubling; resulting plants are diploid and homozygous for all alleles.

11. Cell suspension culture uses rapidly growing calluses cut into small pieces and shaken in a liquid nutrient medium.

a. Single cells or small clumps form a suspension of cells; all produce same chemicals as the plant.

b. This technique is a more efficient way of producing chemicals used in drugs, cosmetics, and agricultural applications than farming plants simply to acquire chemicals they produce.

C. Genetic Engineering of Plants

1. Traditionally hybridization (crossing different varieties or species) was used to produce new plants.

2. Transgenic plants carry foreign genes introduced into their cells.

3. Genetic engineering alters genes of organisms so they have new and different traits.

4. Protoplasts in particular lend themselves to direct genetic engineering in tissue culture.

a. High voltage electric pulses create pores in plasma membrane so new DNA can be introduced.

b. When genes for production of firefly enzyme luciferinase were inserted into tobacco protoplasts, adult plants glowed when sprayed with the substrate luciferin.

c. Foreign DNA can be inserted into a plasmid of *Agrobacterium*; this bacterium infects plant cells and can be used to deliver the recombinant DNA to target cells.

d. John Sanford and Theodore Klein of Cornell University developed a particle gun to bombard plant cells in culture with DNA coated metal particles; later, adult plants are generated.

5. Crops have been engineered to resist frost, fungal and viral infections, insect predation, and herbicides.

6. Future crops could have higher protein content and require less water and fertilizer.

7. Sequencing the genomes of a dicot *Arabidopsis thaliana* and rice will give a

blueprint to the genes of other monocots and dicots.

8. Production of salt-, drought-, and cold-tolerant crops may provide enough food for future populations.

9. Soybeans have been altered to produce healthier fatty acids, and commercial products.

10. Genetic engineering is attempting to improve efficiency of RuBP carboxylase and introduce C4 photosynthesis to rice.

11. Single-gene transfers allow plants to produce human hormones.

a. Corn has made antibodies to deliver radioisotopes to tumor cells.

b. Soybeans make an antibody to treat genital herpes.

c. Researchers can introduce a human gene into tobacco plants using tobacco mosaic virus.

d. Tobacco plants produced antigens to treat non-Hodgkin's lymphoma.

Pollination and Hybridization

Pollination is not difficult with most of the bromeliads, but it does require close attention to the ripening of the stamens and to the pistil during the short period it may receive pollen. This period will differ somewhat among the various species.

With few exceptions, and those principally in the Tillandsias, the flowers of the bromeliads last but a few hours. Other exceptions to this rule may be found in many of the Vriesias. If they blossom in cool weather the flower may carry over to the second day.

Some species of the bromeliads flower early in the morning and may be gone by midday. One species of Aechmea which I found in Brazil baffled me for several days. I finally learned that the petals opened at midnight and by three A. M. those petals were completely disintegrated. By way of contrast the flowers of *Tillandsia crocata* last for four to five days according to the prevailing temperature.

When the pistil and stamens are very short and are far down in a slender corolla of the flower they cannot be seen without carefully removing the sepals and petals so as to expose these reproductory organs. This is the condition in the flowers of *Tillandsia lindenii* or *T. cyanea.*

Some of the giant Hohenbergia species such as *H. Saltzmanii* or *H. Blanchetii,* have very small flowers; the petals may be less than an eighth of an inch long and the flower may be completely filled with nectar. All of these different conditions may be met with if you are interested in the pollination of the bromeliads, but be assured that if you are interested enough to study these different conditions in bromeliads and plant life in general will carry a great and lasting compensation to the better understanding of all life in its myriad manifestations; you will find parallels everywhere.

Once you have determined the time of day that the flower is completely open and the stamens and pistil are in mature form, then you may procure the pollen on a small camel's-hair brush and carry it to the flower of the plant which you may desire to fertilize.

Your problems of pollinating the flowers will vary with different species. In many flowers you will find the stigma of the pistil extending out beyond the stamens. Such a flower is very easy to pollinate, but in many species the stamens may exceed the pistil or may be of even length. In many flowers the stamens may completely surround the stigma so that they would have to be removed in order to pollinate it with the pollen from another flower. If this be the case, then Plant No. 1, the maternal plant, should be emasculated as soon as the flower is open or mature. This means that the stamens must all be removed very care fully so that none of the pollen falls on the stigma to cause self-pollination. The pollen should then be carried to it from Plant No. 2, the paternal plant. This may necessitate a little practice with a steady hand. If you wish to use Plant No. 2 for the same cross, then the same care should be used in removing those stamens. Now the pollen is carried from Plant No. 1 to No. 2 which will make the same cross with the same results, as each flower has the parts of both sexes.

Should the two Species which you wish to cross, mature their flowers at different times of the day, you may find a short period between those two maturity hours that you can pollinate one stigma before it is entirely receptive and the other before it has entirely passed that period and still catch both flowers. But if the time space between the two maturity periods is too great, you may be able to keep the pollen from one plant by taking the stamens and keeping them in waxed paper or in a tightly sealed bottle in the refrigerator for a few hours. I have not been able to keep the pollen of bromeliads for any great length of time, but further experiments may make this a possibility. With some flowers in other families, pollen has been kept for a period from six months to a year.

Should your two subjects be two different true species, then the results of this hybridization will be rather constant. Variation in the offspring will not be very great. However, if one of the parent plants in this cross pollination should be a species and the other a hybrid, or if they both be hybrids, then the offspring will run a most interesting lot of hybrids with the different combinations of character of parents and grandparents.

If you wish to self-pollinate a plant for the propagation of more of the same species, then you carry your brush from flower to flower back and forth from plant to plant, thrusting carefully your brush into each flower, being sure that each stigma is covered with pollen.

Rarely will your bromeliad flowers under cultivation be pollinated unless you do it artificially, as there appear to be few insects that are attracted to them. We believe that most of the pollination of bromeliad flowers in their natural state is performed by humming birds and night moths. In some instances, how ever, the ants and bees have been the pollinators, especially with flowers that carry considerable nectar. The pollen is not dispersed by wind because it is of a semi-moist nature.

Then the rare exception of self-pollination may be found in such species as *Guzmania graminifolia* and other new species of Guzmania which I found in Colombia in 1946. The flower heads on these plants were entirely encased in a jelly-like substance, thus preventing any insect from entering the flower. Also, there are a number of other species that bear flowers which never open, and yet they produce seeds without any outer agent of pollination. In these instances, no doubt, the flowers are sell-pollinated.

Most species of bromeliads have a number of flowers on each inflorescence, but the range in quantity may vary from a single flower, as in *T. usneoides* to as many as 8,000 in *Puya raimoudii* This giant Puya has a flower head twenty feet long and eight feet in circumference which holds these myriad flowers.

Nearly all the bromeliad flowers function as perfect flowers, that is, they contain the reproductory organs of both sexes. However, there are exceptions. The Hechtias have both parts, pistil and stamens, but only one sex can function in each flower. Some plants, considered as male plants, will have flowers in which the stamens are perfectly formed while the pistil is aborted and incapable of receiving pollen. In the female plant the flowers have aborted, undeveloped stamens but have a pistil fully developed and capable of accepting the pollen from another plant. A similar condition may be found with some of the Catopsis and Dyckia species.

With the Cryptanthus species there is still another condition. With very few exceptions, and I have examined hundreds of flowers, I have found only male flowers in the center cluster. These flowers are imperfect and have no trace of female parts, but the flowers in the axils of the bracts below the center cluster are invariably perfect, having both sexes that function. The pollen from the center male flowers is fertile and can be used for pollinating, but these flowers are generally spent before the lower perfect flowers appear. The lower flowers can be self-pollinated or can be used for cross pollination in hybridizing.

Self-pollination by hand is not always successful with some species, as it seems that some of them are self-sterile. This condition is still more prevalent among the hybrids. However, it is often the case that, while self-pollination among the flowers on a single infloresence may not be successful; the pollen from one of those flowers may possibly be successfully used on another similar plant, species or hybrid.

Not always are the results of hybridization an improvement over the parents used, but in most instances they will give the hybridizer an interesting new combination of characters quite worth while. I doubt if anyone can foretell which parent plant will influence the resultant hybrid most. However, after attempting several different crosses by using *Aechmea Weiibachii* as one of the parents, I have as yet to find even one of those crosses to be of any value.

Bigeneric crosses can be most interesting although but few such crosses have survived in horticulture. Some well worth-while Cryptbergias have been developed. The cross between *Cryptanthus beuckerii* and *Biilbergia nutans* has become common. As in many bigeneric crosses, the flowers are aborted and have little attraction, but as a foliage plant it has a definite decorative value.

Until 1930 little had been done in hybridization among the bromeliads excepting in the genera of Cryptanthus, Billbergia and Vriesea, and most of that hybridization was carried on in Europe. Even up until the last ten years, little had been done outside of Europe along this line except in Billbergia and Cryptanthus by Florida and California hybridizers.

Now with all the new species that have been introduced into horticulture in the past twelve years, the possibilities have been greatly enlarged with a most interesting field at hand for horticulturists to enjoy the privilege and experience of assisting in the creation of new, decorative forms and color combinations in the Bromeliaceae.

Evolution and Diversity of Plants

Evolutionary History of Plants

A. Characteristics of Plants (kingdom Plantae)

1. Plants are multicellular eukaryotes with well-developed tissues.
2. Plants live in a wide variety of terrestrial environments.
   a. Land existence is an advantage to photosynthesis; water filters much light.
   b. Carbon dioxide and oxygen are in higher concentrations and diffuse more rapidly in air.
3. Land plants must have adaptations to reduce the loss of water.
   a. Leaves and stems are covered by a waxy cuticle that holds in water.
   b. The leaves have openings (stomates) that open and close to regulate gas and water exchange.

4. All plants protect the embryo from desiccation and some protect their entire gametophyte generations.

5. In some plants, pollen grains are transported by wind or animals to the egg, and the embryo.

B. Relationship to Algae

1. Plants are believed to be closely related to green algae.

   a. Both utilize chlorophylls *a* and *b*.
   b. In both, the food reserve is starch.
   c. The cell walls of both contains cellulose.
   d. Both form a cell plate during cell division.

2. In contrast to green algae, plants have sex organs with an outer layer of non-reproductive cells.

3. Plants have a two-generation life cycle called alternation of generations.

   a. Sporophyte generation is a diploid generation producing haploid spores by meiotic cell division.
   b. Gametophyte generation is a haploid generation producing haploid gametes by mitotic division.

Nonvascular Plants

1. Plants are divided into two main groups: nonvascular and vascular plants.

2. Nonvascular plants include:

   a. hornworts (division Anthocerotophyta)
   b. liverworts (division Hepatophyta)
   c. mosses (division Bryophyta)

3. Nonvascular plants lack specialized tissues for transporting water, minerals, and organic nutrients.

4. They lack true roots, stems, and leaves, although they have root-like, stem-like, or leaf-like structures.

5. The gametophyte is the dominant (most conspicuous) generation.

   a. Flagellated sperm swim to the vicinity of the egg in a continuous film of water.

b. The sporophyte is attached to and derives nourishment from the photosynthetic gametophyte.

6. Nonvascular plants are quite small; the largest being is no more than 20 cm tall.

a. Because sexual reproduction involves flagellated sperm, they are usually found in moist habitats.

b. Mosses compete well in harsh environments because the gametophyte can reproduce asexually.

7. Mosses can dry up; later, when water is available, they photosynthesize again.

8. The three divisions are individual lines of descent; mosses are more closely related to vascular plants.

C. Liverworts

1. Division Hepatophyta contains 10,000 species of liverworts.

2. This name arose in ninth century when it was seen as similar to lobes of the liver.

3. *Marchantia* is a example of this group.

a. It has a flat, lobed thallus about a centimeter in length.

b. The upper surface of thallus is smooth; lower surface bears numerous rhizoids projecting into soil.

c. It reproduces asexually and sexually.

4. Rhizoids are the root-like hair that anchors a bryophyte and absorbs water and minerals from the soil.

5. Asexual reproduction involves gemmae in gemmae cups on upper surface of the thallus.

6. Sexual reproduction depends on antheridia and archegonia.

a. Antheridia are on disk-headed stalks and produce flagellated sperm.

b. Archegonia are on umbrella-headed stalks and produce eggs.

c. Zygote develops into a tiny sporophyte composed of foot, short stalk, and capsule.

d. Spores produced within the capsule of the gametophyte are disseminated by wind.

D. Mosses

1. About 12,000 species of mosses are in the division Bryophyta.

2. Mosses are found in the Arctic through the tropics to parts of the Antarctic.

3. Moss prefers damp, shaded localities; some survive in deserts, others in bogs and streams.

4. Mosses store much water; when they dry out, they become dormant; when it rains, become green.

5. Copper mosses live only in the vicinity of copper and serve as an indicator of ore deposits.

6. Luminous moss lives in caves; cells shaped like lenses focus light on chloroplast grana.

7. Some "mosses" are not true mosses:

   a. Irish moss is an edible alga.
   b. Reindeer moss is a lichen.
   c. Club mosses are vascular plants.
   d. Spanish moss, which hangs from trees in the southern U.S., is a flowering plant.

8. Most mosses can reproduce asexually by fragmentation.

9. Life cycle begins with alga-like protonema developing from germination of a haploid spore.

   a. Three days of favorable growing conditions produce upright shoots covered with leafy structures.
      1) Rhizoids anchor the protonema, to which the shoots are attached.
      2) The shoots bear antheridia and archegonia at their tips.
      3) Antheridia produce flagellated sperm which need external water to reach eggs in archegonia.
      4) Fertilization results in a diploid zygote that undergoes mitotic division to develop a sporophyte.
   b. The sporophyte consists of a foot (which grows down into the gametophyte tissue starting at the former archegonium), a stalk, and an upper capsule (sporangium) where spores are produced.

1) At first the sporophyte is green and photosynthetic.

2) At maturity it is brown and nonphotosynthetic.

E. Uses of Nonvascular Plants

1. *Sphagnum* (bog or peat moss) has tremendous ability to absorb water and is important in gardening.

2. *Sphagnum* does not decay in some acidic bogs; the dried peat can be used as fuel.

F. Adaptation of Nonvascular Plants

1. Nonvascular plants are limited by lack of structural vascular tissue and need for water for sperm.

2. They have advantages living on stone walls, etc. and contribute to soil formation.

Vascular Plants

A. Vascular Tissue

1. Xylem is vascular tissue that conducts water and minerals up from the soil.

2. Phloem is vascular tissue that transports organic nutrients from one part of the plant to another.

B. Organs of Vascular Plants

1. Roots absorb water and minerals from the soil.

2. Stems conduct water and minerals to leaves and conduct organic nutrients from leaves to roots.

3. Leaves are adapted to maximize photosynthetic activity; they are covered by a waxy cuticle.

4. Leaves have openings (stomates) that open and close to regulate gas and water exchange.

C. Life Cycle of Vascular Plants

1. The diploid sporophyte generation is dominant; this is the generation that has vascular tissue..

2. If a faulty gene is present, being diploid can mask it with a functional gene.

D. Reproduction in Vascular Plants

1. Seedless vascular plants (ferns and their allies) disperse their species by producing windblown spores.

a. When spores germinate, a large gametophyte develops independent of sporophyte for nutrition.

b. Antheridia of gametophyte release flagellated sperm that swim in a film of water to archegonia of the gametophyte, where fertilization occurs.

2. In seed plants there is a separate microgametophyte (male) and megagametophyte (female).

3. They are dependent on the sporophyte which is fully adapted to land.

a. Microspores develop within a microsporangium as a result of meiotic cell division.

   1) Haploid microspore develop into grains, immature microgametophytes retained in microsporangium.

   2) When released, the pollen grains develop into sperm-bearing microgametophytes.

   3) In evolution of vascular plants, microgametophytes (pollen grains) replaced external swimming sperm.

b. Megaspores develop within a megasporangium as a result of meiotic cell division.

   1) Haploid megaspores develop into egg-bearing megagametophytes, while still retained in an ovule.

   2) The megagametophyte is dependent on the sporophyte, since it and the subsequent embryo are retained within the ovule that develops into the seed.

Seedless Vascular Plants

A. Seedless Vascular Plants

1. The seedless vascular plants include:

a. whisk ferns (division Psilotophyta),

b. club mosses (division Lycopodopyta),

c. horsetails, (division Equisetophyta)

d. and ferns (division Pteridophyta).

2. These divisions are not closely related.

B. Evolutionary History

1. Rhyniophytes were dominant from mid-Silurian period of the Paleozoic era to the mid-Devonian.

2. *Cooksonia* may have been the first vascular plant.

3. *Rhynia*, a rhyniophyte, is better known; it has erect stems that fork repeatedly.

4. The photosynthetic stems, not true leaves or roots, have sporangia at tips; they are attached to a rhizome.

5. The ancestry of ferns and allies can be traced back to the rhyniophytes or closely related forms.

C. Psilophytes

1. Whisk ferns are in *division Psilophyta* and includes *Psilotum*, a plant that resembles *Rhynia*.

2. Whisk ferns occur in Arizona, Texas, Louisiana, and Florida as well as Hawaii and Puerto Rico.

3. Whisk ferns have no leaves or roots; they possess a branched rhizome with rhizoids and a mycorrhizal fungus that helps gather nutrients.

4. Aerial stems with tiny scales fork repeatedly and carry on photosynthesis.

5. Sporangia are located at the ends of short branches.

6. The independent gametophyte is found underground penetrated by a mycorrhizal fungus; it produces flagellated sperm.

D. Club Mosses

1. About 1,000 species of club mosses are in the division Lycopodophyta.

2. Common in temperate woodlands, they are called "ground pine."

3. A branching rhizome sends up aerial stems less than 30 cm tall.

4. Tightly packed, scalelike microphylls cover stems and branches; each contains one strand of vascular tissue.

5. Sporangia are borne on strobili, terminal clusters of leaves.

6. Spores germinate into inconspicuous and independent gametophytes.

7. Most club mosses live in tropics or subtropics as epiphytes, plants that live on trees without harming them.

8. Closely related are *Selaginella* (spike mosses); dry resurrection plant curls up, unfurls when wet.

9. Club mosses can be traced to Devonian period.

10. They are the only living plants that have microphylls; they are indirectly related to ferns.

E. Horsetails

1. Division Equisetophyta contains one genus, *Equisetum*, with 15 species of horsetails.

2. A rhizome produces aerial stems that stand about 1.3 meters tall.

3. Whorls of slender side branches encircle nodes of a stem, resembling a horse's tail.

4. Leaves have a single megaphyll which has many very small strands of vascular tissue.

5. Many have a strobilus at the tip of stems; others send up special buff-colored stems that bear strobili.

6. Spores germinate into inconspicuous and independent gametophytes.

7. Tough, rigid stems have silica in the cell walls; early Americans used them as "scouring rushes."

F. Ferns

1. 12,000 species of ferns belong to division Pteridophyta.

2. Ferns are widespread, and especially abundant in warm, moist tropical regions.

3. Ferns range in size from low-growing moss-like forms to tall trees.

4. Fronds are variable in size and shape.

5. Nearly all fronds first appear as a fiddlehead which unrolls as it grows.

6. Fronds are megaphylls secondarily subdivided into leaflets; they may have evolved by uneven branching.

7. Uses of Ferns

a. Ferns are used heavily as ornamental plants by florists.

b. Fern wood is very decay resistant.

c. Fern medicines are used by natives to stop bleeding after childbirth; also as an expectorant.

8. Adaptation of Ferns

   a. A water-dependent gametophyte lacks vascular tissue and is separate from the sporophyte.
   b. Flagellated sperm require an outside source of water in which to swim to the eggs.
   c. Once established, bracken fern spreads to drier areas by asexual reproduction.

9. Life cycle of a fern can begin with production of spores by meiotic cell division within sporangia, located in sori on underside of leaflets.

   a. Spores are released and disperse largely by wind.
   b. A spore germinates into a prothallus which grows to develop antheridia and archegonia underneath.
   c. Fertilization occurs if water is present; flagellated sperm swim from antheridia to archegonium.
   d. Resulting zygote begins its development inside archegonium but embryo soon outgrows the space.
   e. Sporophyte becomes visible as the first leaf grows above and as roots develop below the prothallus.
   f. Young sporophyte develops a root-bearing rhizome from which fronds project.

Seed Plants and Gymnosperms

A. The Life Cycle of Seed Plants

1. Gymnosperms include the:

   a. conifers (division Pinophyta)
   b. cycads (division Cycadophyta)
   c. ginkgo (division Ginkgophyta)
   d. gnetophytes (division Gnetophyta)

2. There are separate microgametophytes (male) and megagametophytes (female).

3. Microspores develop immature microgametophytes, pollen grains, still retained in a microsporangium.

4. After they are released, pollen grains develop into mature, sperm-bearing microgametophytes.

5. Pollination is the transfer of pollen to the vicinity of the megagametophyte.

6. Sperm is delivered to an egg through a pollen tube; no external water is required for fertilization.

7. The megaspore develops into an egg-bearing megagametophyte while still retained within an ovule.

8. An ovule is the sporophyte structure that holds the megasporangium and then the megagametophyte.

9. After fertilization, the ovule becomes an embryonic plant enclosed within the ovule, which becomes the seed.

10. Both the megagametophytes and microgametophytes are dependent upon the diploid sporophyte.

11. The sporophyte can evolve into diverse forms without any corresponding changes in the gametophyte.

12. Among seed plants, seeds disperse the sporophytes.

a. Seeds are mature ovules containing embryonic sporophyte and stored food enclosed in protective seed coat.
b. Seeds are resistant to adverse conditions: dryness and temperature extremes.
c. A food reserve supports the emerging seedling until it can exist on its own.
d. Survival value of seeds contributes greatly to success of seed plants, and their present dominance.

B. Gymnosperm Diversity

1. Gymnosperms produce naked seeds not enclosed in a fruit but exposed on the surface of sporophylls.

2. Sporophylls are leaves that bear sporangia and are arranged on a cone.

3. Gymnosperms did not begin to flourish until the Mesozoic era.

a. Pangaea had formed and mountain ranges arose producing deserts on the leeward side.
b. Swamps became much drier and a mass extinction occurred; seedless vascular plants nearly vanished.
c. This provided an opportunity for first seed plants (including gymnosperms) to become dominant.

C. Conifers

1. About 550 species of conifers are in division Pinophyta.

2. Conifers are cone-bearing trees and shrubs such as pines, hemlocks, and spruces.

3. Conifers usually have evergreen needle-like leaves well adapted to withstand extremes in climate.

4. Needles have thick cuticle, sunken stomates, and reduced surface area.

5. Conifers are found in nearly all habitats, from equator to subpolar regions, and comprise the taiga.

6. The oldest and largest trees in existence are conifers:

   a. General Sherman tree in California's Sequoia National Park is 84 meters tall, 10 meters in diameter, and weighs 1,385 tons.

   b. Redwoods grow over 90 meters high and exceed 2,000 years old.

   c. Bristlecone pines in the Nevada mountains are over 4,500 years old.

7. The life cycle of pines is typical of conifers.

   a. Sporophyte is dominant and its sporangia are borne in cones.

   b. Two types of cones are pollen cones (small and near the tips of lower branches) and seed cones.

   c. Each scale of a pollen cone has two or more microsporangia on underside.

   d. Within the sporangia, each microsporocyte undergoes meiosis and produces four microspores.

   e. Each microspore develops into a microgametophyte which is the pollen grain.

   f. Each scale of a seed cone has two ovules surrounded by an integument and with an opening at one end.

   g. A megasporangium is within an ovule; a megasporocyte undergoes meiosis producing four megaspores.

   h. One spore develops into a megagametophyte with 2-6 archegonia, each containing a single large egg.

   i. Once a pollen grain is enclosed within the seed cone, it develops a pollen tube that digests its way toward a megagametophyte and discharges two non-flagellated sperm.

   j. Fertilization takes place one year after pollination.

k. The ovule matures and becomes the seed, composed of embryo, reserve food and seed coat..

l. The woody seed cone, opens to release winged seeds in the fall of the second season.

D. Cycads

1. About 100 species of cycads belong to the division Cycadophyta.

2. The trunk is stout and unbranched; the large leaves are compound giving a palm-like appearance.

3. Cycads flourished during the Mesozoic era and probably were food for herbivorous dinosaurs.

4. Today, cycads are found mainly in tropical and subtropical regions.

E. Ginkgo Trees

1. Only one species of ginkgo (maidenhair tree) survives in the division Ginkgophyta.

2. It grows up to 70 feet in height and is covered with forked-veined, fan-shapod leaves.

3. Microsporangia are in cones but the ovules are stalked; female trees produce seeds with a fleshy covering.

4. It was native to China, preserved in ornamental gardens; now male trees are propagated in city parks.

F. Gnetophytes

1. About 70 species of gnetophytes are in the division Gnetophyta.

2. Gnetophytes are divided into three genera and live mainly in the tropics.

3. *Ephedra* is found in U.S. desert regions, and is a many-branched shrub with small, scalelike leaves.

4. *Welwitschia* is found in desert Africa; most of it exists underground and its two enormous leaves may grow for hundreds of years.

G. Uses of Gymnosperms

1. Gymnosperms supply wood used for building construction and paper production.

2. They produce many valuable chemicals extracted from resin.

H. Adaptation of Gymnosperms

1. Gymnosperms withstand heat, dryness, and cold, as a result of having well-developed roots and stems,

tough, small needles with a thick cuticle.

2. Pollen production has eliminated reliance on external water.

3. Enclosure of the dependent megagametophyte in an ovule protects it during its development.

4. Embryo is protected within seed and is provided nutrients that support growth following germination.

Angiosperms

A. Angiosperms Are Flowering Plants

1. Over 235,000 species of angiosperms (flowering plants) belong to division Magnoliophyta.

2. Angiosperms produce seeds enclosed in fruit.

3. This group contains six times the number of species of all other plants combined.

4. Angiosperms live in all habitats from freshwater to desert and from tropics to subpolar regions.

5. Flowering plant size ranges from microscopic duckweed to Eucalyptus exceeding 100 m tall.

6. They are important in everyday human life: clothing, food, medicine, and commercial products.

B. Origin and Evolution of Angiosperms

1. Angiosperms evolved from ancient gymnosperms.

2. By the Jurassic period in middle of the Mesozoic era, many gymnosperms had features similar to angiosperms.

a. Some had vascular tissue resembling that of angiosperms.

b. Some had sporophylls that resemble early flower parts and were visited by beetles, early insects.

c. Transitional gymnosperms became extinct; the gnetophyte *Ephedra* is considered closest relative.

3. Earliest angiosperms arose in Cretaceous; in Cenozoic, angiosperms diversified as climate grew cold.

4. Rather than being tall trees, the first angiosperms may have been fast-growing woody shrubs.

5. Today angiosperms are either woody or herbaceous.

6. Angiosperms adapted to a wide range of climatic zones are perennials or annuals.

a. Perennials live two or more growing seasons; they die back seasonally in herbaceous plants.

b. Annuals are plants that live for only one growing season.

C. Classification of Flowering Plants

1. Angiosperms are divided into two groups: dicotyledons and monocotyledons.

2. Dicotyledons are in class Magnoliopsida and have these features:

a. either woody or herbaceous,

b. flower parts usually in fours and fives,

c. leaves usually net-veined,

d. vascular bundles arranged in a circle within the stem, and

e. produce two cotyledons (seed leaves) at germination.

3. Dicots include buttercup, mustard, maple, cactus, pea, and rose families (and many more).

4. Monocotyledons are in the class Liliopsida and have these features:

a. most are herbaceous,

b. flower parts are in threes,

c. leaves are usually parallel-veined,

d. vascular bundles are scattered within the stem, and

e. produce one cotyledon (seed leaf) at germination.

5. Monocots include lily, palm, orchid, iris, and grass families (grasses include corn, rice, wheat, etc.).

D. The Flower

1. Flowers have several kinds of highly modified leaves arranged in rings and attached to a receptacle.

2. Receptacle is a modified stem tip to which flower parts are attached.

3. Sepals are outer ring of modified leaves of flowers; usually green, they enclose flower before it opens.

4. Petals are a ring of modified leaves inside of sepals; large and colorful, they help attract pollinators.

5. Stamens form a whorl inside the petals and around a pistil; slender filament has an anther at its tip.

6. Anther is a modified sporophyll.

   a. Anther contains microsporangia where microspores (pollen) are produced.
   b. Anthers likely evolved as a reduction of a sporangium's leaf-like portion.

7. The pistil contains one or more fused carpels; it consists of a stigma, style, and ovary.

   a. Carpels are modified sporophylls that contain ovules in which megasporangia are located.
   b. A stigma is a landing platform for pollen and the site where the pollen tube enters the style.
   c. The style is a slender column that extend from upper part of a carpel and terminates in a stigma.
   d. Ovary contains one to many ovules which develop(s) into seeds; the ovary develops into a fruit.
   e. Fruit is a mature ovary and provides a fleshy or dry covering for seeds; may aid in seed dispersal.

E. Life Cycle of a Flowering Plant

1. Microsporangia develop within the anther portion of a stamen; the megasporangia develop within ovules.

2. Pollination brings mature microgametophyte (pollen grain) to pistil and pollen tube brings sperm to the ovule.

3. Double fertilization: one sperm fertilizes egg; One sperm unites with polar nuclei to form triploid endosperm.

4. Endosperm is contained within a seed as a nutritive tissue that surrounds and serves as food for embryo.

5. The ovule develops into the seed, and the ovary becomes the fruit.

6. Different kinds of fruit employ different kinds of dispersal mechanisms (i.e., wind, gravity, water, and animals) for dissemination of seed(s), which thereby promotes dispersal of the species to new locations.

7. Under favorable conditions, the seed germinates to produce a young sporophyte plant.

F. Flowers and Diversification

1. Flower variety is related to the numerous means by which flowers are pollinated and fruits dispersed.

2. Although some flowers disperse pollen by wind, many attract specific pollinators (e.g., bees, wasps, flies, butterflies, moths, and even bats) which carry only a particular pollen.

3. Co-evolution of pollinators and flowers increases probability flower is pollinated by its own species.

4. Flowers promote efficient cross pollination; they also aid in dispersal through production of fruits.

5. There are fruits that utilize wind, gravity, water, and animals for dispersal.

6. Since animals live in certain habitats or have particular migration patterns, they can deliver a fruit-enclosed seed to a suitable location for germination and development of the adult sporophyte.

G. Uses of Angiosperms

1. Angiosperms are a major food source for humans.

2. Humans use plant fibers to produce cloth, firewood, and construction materials.

3. Plant oils, spices, and drugs are derived from different parts of angiosperm plants.

H. Adaptations of Angiosperms

1. Angiosperms have true roots, stems, and leaves.

2. The vascular tissue is well developed, and leaves are generally broad.

3. The reproductive organs of angiosperms are in their flowers which often attract pollinators.

4. Flowers also produce seeds enclosed by fruit, which aids in dispersal.

Structure and Organization of Plants

Plant Organs

A. Flowering Plant Structure

1. Structures of flowering plants are well-adapted to varied environments.

2. Flowering plants have three vegetative organs: root, stem and leaf.

B. Roots

1. We speak of "a root" but it is more appropriate to call it a root system.

2. Root system is the main root plus its lateral (side) branches.

3. It is generally equal in size to the shoot system, the part above round.

4. Root systems have the following functions.

   a. Roots anchor a plant in soil and give support.
   b. Roots absorb water and minerals from soil; root hairs are central of this process.
      1) Root hair cells are in a zone near root tip.
      2) Root hairs are numerous to increase absorptive surface of a root.
      3) Transplanting plants without preserving root hairs damages a plant.
      4) Water and nutrients absorbed are distributed to rest of the plant.
      5) Roots produce hormones that must be distributed to the plant
   c. Perennials "die back" to regrow next season; roots store food. (e.g., carrots, sweet potatoes).

C. Stems

1. Stem forms main axis of a plant, along with lateral branches.

2. Stem produces leaves and arrays them to be exposed to as much sun as possible..

3. A node occurs where a leaf attaches and internode is region between nodes; nodes and internodes identify a stem even if it is underground.

4. Stem has vascular tissue to transport water and minerals from roots and sugar from leaves.

5. Non-living cells form a continuous pipeline in vascular tissue.

6. A cylindrical stem expands in girth and length; trees use woody tissue to strengthen stems.

7. Stems function in storage: cactus stems store water; tubers are horizontal stems that store nutrients.

D. Leaves

1. A foliage leaf is the usual organ of photosynthesis in vascular plants.

2. Leaves receive water from roots by way of the stem.

3. Broad, thin leaves have maximum surface area to absorb CO2 and collect solar energy.

4. A blade is the wide portion of a leaf with most photosynthetic tissue.

5. Petiole is a stalk that attaches a leaf blade to stem.

6. Leaf axil is upper acute angle between petiole and stem where an axillary (lateral) bud originates.

7 Some leaves protect buds, attach to objects, store food, or capture insects.

Monocot Versus Dicot Plants

A. Criteria for Monocots and Dicots

1. Flowering plants are divided into monocots and dicots based on these traits.

a. Number of cotyledons in seed one two

b. Distribution of root xylem and phloem root xylem and phloem in a ring root phloem between arms of xylem

c. Distribution of vascular bundles scattered in stem arranged in a distinct ring

d. Pattern of leaf veins form a parallel pattern form a net pattern

e. Number of flower parts in threes and multiples of threes in fours, fives and multiples of fours or fives

f. Number of apertures in pollen usually one usually three

2. Representative members: grasses, lilies, orchids dandelions to oak trees and palm trees

3. Cotyledons are embryonic seed leaves providing nutrition before mature leaves begin photosynthesis.

4. These traits represent an evolutionary path dating back to origin of flowering plants.

Plant Tissues

A. Meristem Produces Tissue

1. Plants continually grow due to meristem (embryonic tissue) in stem and root tips.

2. Three types of primary meristem produce three types of specialized tissue.

   a. Protoderm is outermost primary meristem giving rise to epidermis.
   b. Ground meristem is inner meristem producing ground tissue.
   c. Procambium produces vascular tissue.

3. Three specialized tissues are produced.

   a. Epidermal tissue forms outer protective covering.
   b. Ground tissue fills the interior.
   c. Vascular tissue transports water and nutrients and provides support.

B. Epidermal Tissue

1. Epidermis is an outer protective covering tissue of plant roots, leaves, and stems of nonwoody plants.

2. It contains closely packed epidermal cells.

3. Waxy cuticle covers walls of epidermal cells, minimizing water loss and protecting against bacteria.

4. In roots, certain epidermal cells are modified into root hairs that increase surface area of the root for absorption of water and minerals and help to anchor plants.

5. Different protective hairs are produced by epidermal cells of stems and leaves.

6. Epidermal cells are modified as glands to secrete protective substances.

7. On lower epidermis of leaves, special guard cells form microscopic pores (stomates) and regulate gas exchange between leaf interior and exterior.

8. In older woody plants, epidermis of the stem is replaced by cork tissue.

   a. Cork is outer covering of the bark of trees; composed of dead cork cells that may be sloughed off.
   b. Cork cambium is lateral meristem that produces new cork cells.
   c. As cork cells mature, they encrust with lipid suberin that renders them waterproof and inert.

d. Cork protects a plant and makes it resistant to attack by fungi, bacteria, and animals.

C. Ground Tissue

1. Ground tissue fills the inside of plants with parenchyma, collenchyma and sclerenchyma cells.

2. Parenchyma are least specialized of all plant cell types.

   a. Cells of this type contain plastids (e.g., chloroplasts or colorless storage plastids).
   b. They are found in all organs of a plant.
   c. They divide to form more specialized cells (e.g. roots develop from stem cuttings in water).

3. Collenchyma resemble parenchyma but has thicker primary cell walls.

   a. Collenchyma cells are uneven in the corners.
   b. They usually occur as bundles of cells just beneath epidermis.
   c. They give flexible support to immature regions of plants (e.g. celery stalk is mostly collenchyma).

4. Sclerenchyma cells have thick secondary cell walls.

   a. They are impregnated with lignin that renders walls tough and hard.
   b. They provide strong support to mature regions of plants.
   c. Most cells of this type are non-living.
   d. Sclerenchyma cells form fibers (used in linen and rope) and shorter sclereids (found in seed coats and nut shells).

D. Vascular Tissue

1. Xylem passively conducts water and mineral solutes upward through a plant from roots to leaves;

   a. Xylem contains tracheids, vessel elements, and parenchyma cells that store various substances.
   b. Tracheids
      1) Tracheids are hollow, thin, long non-living cells with tapered overlapping ends.
      2) Water moves across end and sidewalls because of pits or depressions in secondary cell wall.

c. Vessel Elements

1) Vessel elements are hollow non-living cells lacking tapered ends.

2) They are larger than tracheids.

3) They lack transverse end walls.

4) They form a continuous pipeline for water and mineral transport.

d. Xylem also contains sclerenchyma cells to add support.

2. Phloem is vascular tissue that conducts the organic solutes in plants, from the leaves to the roots; it

contains sieve-tube cells and companion cells.

a. Sieve-tube cells

1) Sieve-tube cells contain cytoplasm but no nucleus.

2) They are arranged end to end.

3) They have channels in their end walls (thus, the name "sieve-tube"), through which plasmodesmata extend from one cell to another.

b. Companion Cells

1) Companion cells are closely connected to sieve-tube cells by numerous plasmodesmata.

2) They are smaller and more generalized than sieve-tube cells.

3) They have a nucleus which may control and maintain the function of both cells.

4) They are also thought to be involved in the transport function of phloem.

3. Vascular tissue extends from root to leaves as vascular cylinder (roots), vascular bundles (stem) and leaf veins.

Organization of Roots

A. Dicot Root Tip

1. Dicot root tip, site of primary growth, is organized into zones of cells in various stages of differentiation.

2. Cells are continuously added to a root cap below and zone of elongation above by contributions from

the zone of cell division.

3. Root cap is a protective cover; its cells are replaced constantly because they are soon ground off.

4. Zone of elongation is above a zone of cell division where cells become longer and more specialized.

5. Zone of cell division contains meristematic tissue and adds cells to root tip and zone of elongation.

6. Zone of maturation is above zone of elongation; cells are mature and differentiated and it has root hairs.

B. Tissues of a Dicot Root

1. Epidermis is a single layer of thin-walled, rectangular cells.

   a. Epidermis forms protective outer layer of the root.
   b. In region of maturation, there are many root hairs.
   c. Root hairs project as far as 5-8 mm into the soil.

2. Cortex is a layer of large, thin-walled, irregularly shaped parenchyma cells.

   a. These cells contain starch granules; function in food storage.
   b. Cells are loosely packed; water and minerals can diffuse through cortex without entering cells.

3. Endodermis is layer of rectangular cells; forms boundary between cortex and inner vascular cylinder.

   a. Its cells fit closely together and are bordered on four sides by the Casparian strip.
   b. It regulates entrance of minerals into vascular cylinder.
   c. Casparian strip is impermeable lignin and suberin layer that excludes water and mineral ions.
   d. The only access to the vascular bundle is through endodermal cells.

4. Vascular cylinder is an arrangement of vascular tissues as a cylinder.

   a. Pericycle is first layer of cells within vascular cylinder
      1) Its cells have retained capacity to divide.
      2) It can start development of branch or secondary roots.
   b. Vascular tissue forms main portion of a vascular bundle; it is composed of
      1) xylem, whose cells are arranged in a star-shaped pattern; and
      2) phloem, whose cells are located in regions between arms of xylem.

C. Organization of Monocot Roots

1. Monocot roots do not undergo secondary growth.

2. Monocot root has a ring of vascular tissue; alternating bundles of xylem and phloem surround pith.

3. Monocot roots also have pericycle, endodermis, cortex, and epidermis.

D. Root Diversity

1. Roots have adaptations to help anchor plants, absorb water and minerals, and store carbohydrates.

2. There are three general root types:

   a. Taproot is common in dicots; first or primary root grows straight down and remains dominant root of a plant; often fleshy and adapted to store food (e.g., carrots, beets).
   b. Fibrous root system of monocots is a mass of slender roots and lateral branches that hold soil.
   c. Adventitious roots develop from underground stems or from base of above-ground stems.

      1) Prop roots main function is to anchor a plant (e.g. corn and mangrove plants).
      2) Pneumatophores of mangrove plants project above the water from roots to acquire oxygen.
      3) Ivy has holdfast roots to anchor aerial shoots.

3. Haustoria are root-like projections of stems of parasitic plants (e.g., dodders and broomrapes).

   a. They grow into the host plant.
   b. They contact vascular tissue from which they extract water and nutrients.

4. Mycorrhizae are fungus roots.

   a. In this mutualism, fungus receives sugars and amino acids from plant.
   b. Plant receives water and minerals from the fungus.

5. Legumes (e.g., peas and beans) have root nodules containing nitrogen-fixing bacteria.

a. Bacteria extract nitrogen from air and reduce it to a form that can be used by plant tissues.

b. Legumes are often planted to bolster nitrogen supply of soil.

Organization of Stems

A. Primary Growth

1. Stem tip is site of primary growth where cell division extends length of stems or roots.

2. Shoot apical meristem produces new leaves and primary meristems; increases stem length.

3. Shoot apical meristem is protected within a terminal bud of leaf primordia (immature leaves).

4. Bud scales are scale-like coverings protecting terminal buds during winters when bud growth stops.

5. Three specialized types of primary meristem develop from shoot apical meristem.

   a. Protoderm is outermost primary meristem that gives rise to epidermis.

   b. Ground meristem produces two tissues composed of parenchyma cells: pith and cortex.

   c. Procambium is inner meristem that produces primary xylem and primary phloem.

6. Differentiation continues; cells become first tracheids or vessel elements within vascular bundle.

7. First sieve-tube cells are short-lived and do not have companion cells.

8. Mature phloem develops later after all surrounding cells have stopped expanding.

B. Herbaceous Stems

1. Herbaceous stems are mature nonwoody stems that exhibit only primary growth.

2. Outermost tissue of herbaceous stems is epidermis covered by waxy cuticle to prevent water loss.

3. Xylem and phloem are in distinctive vascular bundles.

   a. In each bundle, xylem is found to inside of stem; phloem is found to outside.

   b. In dicot herbaceous stem, vascular bundles are arranged in a ring towards outside of the stem and separating cortex from central pith.

c. In monocot stem, vascular bundles are scattered throughout the stem; there is no well-defined cortex or pith.

4. Cortex sometimes carries on photosynthesis; pith may function as a storage site.

C. Woody Stems

1. Woody plants have both primary and secondary tissues.

2. Primary tissues are new and form each year from primary meristem right behind apical meristem.

3. Secondary tissues develop from second year onward from growth of lateral meristem.

4. Primary growth increases length of a plant; secondary growth increases its girth.

5. As secondary growth continues, it is not possible to distinguish individual vascular bundles.

6. Woody dicot stem has a different organization with three distinct areas: bark, wood, and pith.

7. Rays are of living cells that allow materials to move laterally.

8. Bark of a tree contains cork, cork cambium, and phloem.

a. Secondary phloem is produced each year by vascular cambium but does not build up.

b. This phloem tissue is soft; therefore it is easy to remove the bark of a tree..

9. Cork cambium is a lateral meristem that produces new cork cells when needed.

a. Cork cambium begins to divide, producing cork that disrupts epidermis replacing it with cork cells.

b. Cork cells become impregnated with suberin, causing them to die but making them waterproof.

c. Consequently, cork forms an impervious barrier, even to gas exchange, except at lenticels,

10. First flowering plants were probably woody shrubs; herbaceous plants evolved later.

a. It is advantageous to be woody when there is adequate rainfall; woody plants can grow taller and have adequate tissue to support and service leaves.

b. It takes energy to support secondary growth and prepare plant for winter in temperate zones.

c. Long-lasting plants need more defense mechanisms against attack by herbivores and parasites.

d. Trees need years to mature before reproducing; they are more vulnerable to accident or disease.

11. Annual Rings

a. Vascular cambium is dormant during winter.

b. Spring wood is composed of wide xylem vessel elements with thin walls, necessary to conduct of sufficient water and nutrients to supply abundant growth that occurs during spring.

c. Summer wood forms when moisture is scarce; composed of a lower proportion of vessels, it contains thick-walled tracheids and numerous fibers.

d. Annual ring is one ring of spring wood followed by a ring of summer wood; equals one year's growth.

e. Sapwood is outer annual rings where transport occurs.

f. Heartwood is inner annual rings of older trees.

1) Vessels no longer function in transport; they become plugged with resins and gums that inhibit growth of bacteria and fungi.

2) Heartwood may help to support a tree.

D. Stem Diversity

1. Stolons are stems that grow along ground; new plants grow where nodes contact soil.

2. Succulent stems of cacti are modified for water storage.

3. Tendrils of grapes and morning glories are stems adapted for wrapping around support structures.

4. Rhizomes are underground horizontal stems.

a. Rhizomes are long and thin in grasses and thick and fleshy in irises.

b. Rhizomes survive winter and contribute to asexual reproduction because each node bears a bud.

c. Some rhizomes have tubers that function in food storage (e.g., potatoes).

5. Corms are bulbous underground stems that lie dormant during winter, like rhizomes.

6. Humans use stems: sugarcane is primary source of table sugar; cinnamon and quinine are from bark.

Organization of Leaves

A. Leaf Structure

1. Leaves are organs of photosynthesis in plants; they are made of a flattened blade and a petiole.

2. Leaf veins reveal the presence of vascular tissue within the leaves.

3. The vascular tissues of leaves transport water and nutrients.

4. Leaf veins have a net pattern in dicot leaves and a parallel pattern in monocot leaves.

5. A petiole is a stalk that attaches a leaf blade to the plant stem.

6. Epidermis is the layer of cells that covers the top and bottom sides of a leaf.

   a. Epidermis often bears protective hairs or glands; epidermal glands produce irritating substances.

   b. Epidermis is covered by a waxy cuticle that keeps the leaf from drying out.

   c. Epidermis, particularly lower epidermis, contains stomates that allow gases to move into and out of leaf.

7. Mesophyll is the inner body of a leaf and the site of most of photosynthesis.

   a. Palisade mesophyll is layer of mesophyll containing elongated parenchyma cells with many chloroplasts.

   b. Spongy mesophyll contains loosely packed parenchyma cells that increase surface area for gas exchange.

B. Leaf Diversity

1. Simple leaves have margins not deeply lobed or divided into smaller leaflets.

2. Compound leaves are divided into smaller leaflets, and each leaflet may have its own stalk.

3. Leaves are variously modified.

   a. Cactus spines are modified leaves; succulents have fleshy leaves to hold moisture.
   b. Onion bulbs have leaves surrounding a short stem.
   c. Tendrils of peas and cucumbers are leaves.
   d. Venus's-flytrap has leaves to trap and digest insects.

Control of Growth and Responses in Plants

Plant Responses

A. Organisms Respond to Stimuli

1. One defining characteristicoflife is an ability to respond to stimuli.

2. Adaptive organisms respond to environmental stimuli because it leads to longevity and survival.

3. Animals have nerves and muscles; plants respond by growth patterns.

B. Tropisms

1. A tropism is plant growth toward or away from a directicnal stimulus.

   a. Directional is important; the stimulus comes from only one direction instead of many.
   b. Growth toward a stimulus is a positive tropism; growth away from a stimulus is a negative tropism.
   c. By differential growth, one side elongates faster; result is curving toward or away from a stimulus.

2. Three well-known tropisms are named for the stimulus that causes the response.

   a. Phototropism is growth of plants in response to light; stems show positive phototropism.
   b. Gravitropism is response to earth's gravity; roots demonstrate positive gravitropism and stems demonstrate negative gravitropism.
   c. Thigmotropism is unequal growth due to touch (e.g., coiling of tendrils around a pole).

C. Phototropism

1. Early researchers, including Charles Darwin and his son, observed plants curve toward light.

2. Phototropism occurs because cells on shady side of stems elongate.

3. It is believed that a yellow pigment related to riboflavin acts as a photoreceptor for light.

   a. Following reception, plant hormone auxin migrates from bright side to shady side of a stem.

   b. How reception of stimulus couples to production of auxin is not yet known.

4. Auxin is also involved in gravitropism, apical dominance, and root and seed development.

D. Gravitropism

1. An upright plant placed on its side displays negative gravitropism; it grows upward opposite gravity.

2. Charles Darwin and his son were first to show that roots display positive gravitropism.

   a. If root cap is removed, roots no longer respond to gravity.

   b. Later researchers showed root cap cells contain statoliths, starch grains within amyloplasts.

   c. Due to gravity, amyloplasts settle to lowest part of cell.

3. The hormone auxin underlies both positive and negative gravitropisms.

   a. The two tissues respond differently to auxin, which moves to lower side of both stems and root.

   b. Auxin inhibits growth of root cells; cells of upper surface elongate and root curves downward.

   c. Auxin stimulates growth of stem cells; cells of lower surface elongate and stem curves upward.

E. Thigmotropism

1. Unequal growth due to contact with solid objects is thigmotropism.

2. Coiling of morning glory or pea tendrils around posts, etc. is a common example.

3. Cells in contact with an object grow less while those on opposite side elongate.

4. This process is quite rapid; tendrils have been observed to encircle an object in ten minutes.

5. A couple of minutes of stroking can bring about a response that lasts for several days.

6. Response can be delayed; tendrils touched in the dark will respond when illuminated.

   a. ATP rather than light can cause the response; need for light is a need for ATP.
   b. Hormones auxin and ethylene are involved; they induce curvature of tendrils in absence of touch.

7. Thigmomorphogenesis is a touch response involving the whole plant.

   a. Entire plant responds to presence of wind or rain.
   b. A plant growing in a windy location has a shorter, thicker trunk.
   c. Even rubbing of a plant inhibits cellular elongation and produces a shorter, sturdier plant.

F. Nastic Movements

1. In contrast to tropisms, nastic movements are independent of the direction of stimulus.

2. Seismonastic movements result from touch, shaking, or thermal stimulation.

3. This response takes only a second or two, is due to a loss of turgor pressure within cells.

4. When a *Mimosa pudica* leaf is touched, leaflets fold because petiole droops.

5. A pulvinus is a thickening at base of such leaflets where turgor pressure can rapidly drop.

6. Mechanisms are potassium ions that move out of the cell; water follows by osmosis.

7. A single stimulus such as a hot needle can cause leaves to respond.

8. Venus's-flytrap

   a. This plant has three sensitive hairs at base of the trap.
   b. When touched by an insect, an impulse-type stimulus triggers the trap to close.
   c. Turgor pressure in leaf cells propel the trap.

G. Sleep Movements

1. Sleep movements are nastic responses to daily changes in light level.

2. Movement is due to changes in turgor pressure of motor cells in a pulvinus.

3. Some plant movements correspond to environmental changes in light, temperature, etc.

4. Circadian rhythm is a biological rhythm with a 24-hour cycle.

5. Biological clock is an internal mechanism maintaining biological rhythms in absence of stimuli.

6. Biological clocks are synchronized by external stimuli to twenty-four-hour rhythms.

7. Photoperiod is more reliable an indicator of seasonal changes than temperature change.

8. Stomates and flowers usually open in morning, close at night; some plants secrete nectar at same time of day.

Plant Hormones

A. For Plants to Respond to Stimuli, Activities of Plant Cells and Structures have to be Coordinated.

1. Almost all plant communication is done by hormones.

2. Hormones are low concentration chemical messengers active in another part of an organism.

3. A responses is influenced by several hormones; requires a specific ratio of two or more hormones.

4. Hormones are synthesized in one part of a plant; they travel in phloem after plant receives appropriate stimulus.

5. Each naturally-occurring hormone has a specific chemical structure.

6. Other chemicals that differ only slightly from natural hormones also affect the growth of plants.

7. Plant growth regulators are hormone imitators plus naturally-occurring plant growth hormones.

B. Auxin

1. Indoleacetic acid (IAA) is most common naturally-occurring auxin.

   a. It is produced in shoot apical meristem and found in young leaves, flowers, and fruits.

b. Apically-produced IAA prevents growth of axillary buds; this provides for apical dominance.

c. When a terminal bud is removed, nearest buds begin to grow and the plant branches.

d. Application of a weak solution of auxin causes roots to develop from ends of cuttings.

e. Auxin production by seeds promotes growth of fruit.

f. As long as auxin is concentrated in leaves and fruits rather than stem, they do not fall off.

2. Auxin-controlled cell elongation is involved in gravitropism and phototropism.

a. When gravity is perceived, auxin moves to lower surface of roots and stems.

b. Darwin discovered with oat seedlings, phototropism would not occur if tip of a seedling is cut off or covered by a cap; they concluded cause of curvature moved from coleoptile tip to rest of shoot.

3. Frits W. Went experimented with coleoptiles in 1926.

a. He cut off tips and placed them on agar.

b. Agar block was placed to one side; coleoptile would curve away from that side regardless of light.

c. He deduced a chemical caused curved growth and named it auxin after Greek word for "to grow."

C. How Auxins Work

1. In a plant exposed to unidirectional light, auxin moves from bright side to shady side of a stem.

2. Auxin binds to receptors and activates the ATP-driven proton (H+) pump.

3. As hydrogen ions are pumped out of the cell, cell wall becomes acidic, breaking hydrogen bonds.

4. Cellulose fibrils are weakened and activated enzymes further degrade the cell wall.

5. Electrochemical gradient established causes uptake of solutes; water follows by osmosis.

6. Turgid cell presses against cell wall, stretching it so that elongation occurs.

7. Auxin-mediated elongation occurs in younger cells; older cells may lack auxin receptors.

D. Gibberellins

1. Gibberellins are a group of 70 plant hormones that chemically differ.

2. GA3 is the most common of natural gibberellins.

3. Gibberellins are growth promoters that elongate cells.

4. Gibberellins were discovered in 1926 by Ewiti Kurosawa, a Japanese scientist investigating a fungal

disease of rice plants called "foolish seedling disease."

a. His fungus-infected plants produced an excess chemical gibberellin, named after the fungus.

b. By 1956, gibberellic acid was finally isolated from a flowering plant rather than fungus.

5. Mode of action

a. Hormone GA3 binds to a receptor; a second messenger (Ca2+) inside cell combines with the protein calmodulin.

b. Ca2+ -calmodulin complex activates a gene coding for the enzyme amylase.

c. Amylase acts on starch to release sugars used as a source of energy by the growing embryo.

E. Cytokinins

1. Cytokininsare a class of plant hormones that promote cell division.

2. Cytokinins are derivatives of the purine base adenine.

3. A natural cytokinin zeatin is found in corn kernels; kinetin is a synthetic cytokinin.

4. Researchers discovered cytokinins in work on growing plant tissues in culture.

5. Oligosaccharins, chemical fragments released from cell wall, also direct differentiation.

6. Researchers hypothesize that auxin and cytokinins are part of a reception-transduction-response pathway that activates enzymes that release these fragments from cell wall.

F. Senescence

1. Aging processes are senescence; large molecules break down and transported elsewhere.

2. Cytokinins prevent senescence of leaves; they also initiate development of leaf growth.

3. Cytokinins initiate growth of lateral buds despite apical dominance.

G. Inhibitory Hormones

1. Abscisic acid (ABA) is called "stress hormone"; it maintains seed and bud dormancy and causes closure of stomates.

2. Dormancy occurs when a plant organ readies itself for adverse conditions by stopping growth.

a. ABA moves from leaves to vegetative buds in fall; thereafter these buds are converted to winter buds which are covered by thick, hardened scales.

b. Reduction in ABA and increase in gibberellins break seed and bud dormancy; seeds germinate and buds send forth leaves.

3. Abscisic acid brings about closing of stomates when a plant is under water stress.

a. By some unknown mechanism, ABA causes K+ ions to leave guard cells.

b. As a result, guard cells lose water and stomates close.

4. Although external application of ABA promotes abscission, it is not believed to function in this process; the hormone ethylene is considered to have this natural function.

H. Ethylene

1. It was an early practice to prepare citrus fruit for market by storage in a room with a kerosene stove.

2. Later work revealed incomplete combustion of kerosene produced ethylene which ripens fruit.

3. Ethylene is a gaseous plant hormone; it ripens fruit by increasing activity of enzymes that soften fruit.

4. Ethylene stimulates production of cellulase enzyme that hydrolyzes cellulose in plant cell walls.

5. A barrel of ripening apples can induce ripening of a bunch of bananas some distance away.

6. Ethylene releases from site of a physical wound; therefore one rotten apple spoils the whole bunch.

7. The presence of ethylene in air inhibits growth of plants in general.

8. Ethylene is present in auto exhaust and in homes heated with natural gas.

9. Inhibition of plant growth occurs in low concentrations (one part ethylene per 10 million parts of air).

10. Ethylene is involved in abscission, the dropping of leaves, fruits, or flowers.

   a. Lower levels of auxin in these areas (compared to stem) probably initiate abscission.
   b. Once abscission begins, ethylene stimulates production of enzymes such as cellulase that cause leaf, fruit, or flower drop.

Photoperiodism

A. Many physiological changes in plants (e.g. seed germination, the breaking of bud dormancy, flowering, and the onset of senescence) are related to a seasonal change in day length.

1. Photoperiodism is a physiological response to relative lengths of daylight and darkness.

2. Research by U.S. Department of Agriculture in controlled greenhouses revealed this mechanism.

B. Plants can be divided into three groups, based on photoperiodism.

1. Short-day Plants

   a. These flower when day length was shorter than a critical length.
   b. Examples include cockle-bur, poinsettia, and chrysanthemum.
   c. In effect, they require a period of darkness that is longer than a critical length to flower.

2. Long-day Plants

   a. These flower when the day length is longer than a critical length.
   b. Examples include wheat, barley, clover, and spinach.

c. In effect, they require a period of darkness that is shorter than a critical length to flower.

3. Day-neutral Plants

a. These are plants for which flowering is not dependent on day length.

b. Examples include tomato and cucumber.

C. A long-day and a short-day plant can have same critical length.

1. These plants differ in specific sequence of day lengths as a season progresses.

2. Spinach is a long-day plant that flowers in summer when day length increases to 14 hours.

3. Ragweed is a short-day plant that flowers in fall when day length shortens to 14 hours or less.

4. In 1938, K. C. Hammer and J. Bonner experimented with artificial lengths of dark and light periods.

a. Cocklebur, a short-day plant, flowers as long as the dark period lasts over 8.5 hours.

b. If dark period is interrupted by a flash, it does not flower; darkness amidst day cycle has no effect.

c. Long-day plants require a dark period shorter than a critical length regardless of length of light period.

d. Therefore, length of the dark period controls flowering, not length of the light period.

D. Phytochrome and Flowering

1. U.S.D.A. scientists discovered phytochrome, a blue-green leaf pigment that exists in two forms:

2. Pr(phytochrome red) absorbs red light (wavelength of 660; nm); it is converted to Pfr.

3. Pfr is phytochrome far-red and absorbs far-red light (wavelength of 730 ;nm); it is converted to Pr.

4. During a 24-hour period, there is a shift in ratio of these two pigments.

a. Direct sunlight contains more red than far-red light; Pfr is present in plant leaves during the day.

b. Shade and sunsets have more far-red than red light; Pfr is converted to Pr as night approaches.

c. There is a slow metabolic replacement of Pfr by Pr during night.

5. Phytochrome conversion may be a first step in reception-transduction-response pathway resulting in flowering.

E. Other Functions of Phytochrome

1. The Pr ——> Pfr conversion cycle controls other growth functions in plants.

2. In addition to being involved in flowering, Pfr promotes seed germination and stem branching.

3. Following germination, presence of Pr dominates; stem elongates and grows toward sunlight while

leaves remain small.

4. Once a plant is exposed to sunlight and Pr is converted to Pfr, the plant begins to grow normally-leaves expand and the stem branches.

5. Pfr form of phytochrome triggers activation of one or more regulatory proteins in cytosol.

6. These proteins migrate to nucleus and bind to "light-stimulated" genes coding for proteins found in chloroplasts.

5

Agriculture Technology and Food Industry: Tools & Techniques

New Technologies and their Impact on Agriculture, Environment And The Food Industry

The agri-food sector is facing a number of global challenges that require a re-evaluation of current practices in production and trade, the cooperation between enterprises along the vertical supply chain, relationships between enterprises on similar stages of production or trade, the sector's infrastructure in production and services, and the influence of governments on enterprises' management activities. These challenges include but are not limited to increased globalisation and competition, highly differentiated and segmented food production, complex requirements on quality assurance, reliability and flexibility in the provision of food, sustainability in people's trust, control on environmental effects, and efficiency in the sector's organization and processes.

Information technologies (IT) have the potential to support the agri-food sector in coping with the challenges but they are also key enablers for some of the developments to take place. Today's drive towards globalisation builds on modern communication technology, but it is also accelerated by the technology's communication ability. From this dual perspective, the adoption of IT by members of the agri-food sector is no longer a question of choice but of survival. A choice, however, is the extent the sector will utilize the support potential of IT within the not so distant future. However, the difficulty in anticipating technology evolution and its effects is compounded not only by its rapid change and lack of understanding of technology but also because the ultimate evolution is a social choice. Society will ultimately choose a potential outcome depending on decisions made on investment, acceptance, adoption and rejection (Schiefer and Zazueta, 2003).

The evolution builds on decisions by the many groups that constitute the sector's activity, involving enterprises, policy, extension or service institutions of any kind and on their cooperation in the specification of a common development view. It is the objective of the paper to support these processes, and to provide some insight into the issues, opportunities and limitations regarding the potential role of IT in future sector activities.

The implementation of the development paths presented in this paper will transform the sector, the production and trade of food products, the relationships between participants in the supply chain reaching from farmers to consumers and the market infrastructure in agriculture and the food sector. These effects evolve from integrating different IT development lines and their underlying technology components into IT application environments that not only improve present activities but eventually replace the business model of today's agri-food sector.

The Application Environment

Today's agri-food sector has to simultaneously face critical challenges from a variety of sources. Globalization increases competition but also involves higher risks in food safety or quality. This development coincides with increasing pressures on the agri-food sector to intensify process controls and to improve on quality, food safety, the tracking and traceability of products throughout the supply chain, and the environmental consequences of its operations. This combination mounts to an unparalleled challenge regarding the sector's organization and efficiency. None of these challenges can be met by individual enterprises or enterprises of a certain stage in the supply chain as, e.g., farms, alone. The close dependencies between all levels of food production require joint initiatives and new approaches for cooperation. However, while the initiatives require a cooperate approach, they primarily build on changes in enterprises' internal activities and their interaction with each other, IT is the key enabler and efforts are being made to integrate IT opportunities in an appropriate way into these activities. Within these activity categories, the primary focus of current IT developments is on three groups of activities, market activities, process activities, and management decision and extension activities.

Process Activities

Process activities in this context refer to enterprise internal processes in food production and production control. Discussions on IT support for process improvements involve, a.o., process reliability, process control, process efficiency, and the utilization of specific IT developments in comprehensive process management approaches like, e.g., precision agriculture or GPS. The primary focus of IT support is automatic control and its process optimisation.

Market Activities

Market activities of enterprises focus on trading, logistics, and marketing. These activities determine market related business processes and are of relevance for the organization and efficiency of sector operations. Discussions on IT support for market improvements involve, a.o., food quality, food safety, traceability, efficient consumer response (flexibility), transaction efficiency, communication to support consumers' trust, and supply chain cooperation. The keyword for IT support is communication and the utilization of the emerging integrated communication technologies.

Management Decision and Extension Activities

Management decision support through appropriate IT-based systems like MIS (Management Information Systems), DSS (Decision Support Systems) or EIS (Executive Information Systems) housed inside an enterprise or provided by extension is established practice. They involve the collection, selection, processing and communication of information in one or two-way communication activities. Present IT developments add new dimensions to the accessibility and communication of information. They focus on comprehensive IT support environments that integrate with knowledge networks with local, regional or global knowledge bases.

Principal Technology

Development Lines The far-reaching effects of IT on all aspects of society are common knowledge and expressed by references to today's age as the 'information age'. IT refers to a rapidly expanding range of services, methods, techniques, applications, equipment, and electronic technologies used for the collection, manipulation, processing, classification, storage, and retrieval of recordable information and knowledge. At this time, such technologies include, but are not limited to, computers, personal digital assistants (PDAs), software, high-capacity storage, networks, telecommunications, databases, data warehouses, multimedia, and training, the internet and its world wide web, geographic information systems (GIS), computer-aided design (CAD), online services, video conferencing, electronic trade, executive information systems (EIS), electronic mail, and expert systems: in short, all technologies related to the acquisition, storage, recovery, transfer, manipulation, and delivery of data, sound, and graphics, including video.

Any single technology within this almost unlimited variety could be linked to human activities in the agri-food sector and might have a profound effect on them. However, one could delineate groups of related technologies with a similar direction of impact on the sector. They constitute major IT development lines that could directly be linked to future developments of the agri-food sector:

1. Digital integration: eliminates technology breaks
2. Multi-media interaction: utilizes the full potential of human perception
3. Electronic communication networks: provide communication infrastructure
4. Information Portal technology: provides access points to digital knowledge spheres

5. Virtual platforms for collaboration: facilitates digital group interaction
6. Agent technology: reduces needs for human interventio

They all build on the ongoing digital integration that allows uninterrupted information flows from the source to the end and on the emergence of the multi-dimensional information sphere that builds on internet information technology and defines an digital information environment in its own rights. It allows, within its sphere, the creation of all types of communication infrastructures, communities, warehouses, shops, meeting places, services, etc., i.e., a digital duplication of our visible world.

The development lines are related to all areas of sector activities and change the way they are being performed in the future. However, the potential impact of IT on the sector as a whole is most pronounced in market activities which determine the sector's infrastructure, the interaction of enterprises and the transaction of food products on the regional, national or global level. The following discussion will, therefore, concentrate on market related IT applications that provide the basic enablers for major sector developments.

Market Focused Technologies

Market technologies reach from basic document exchange technologies to highly sophisticated support for complex food trading activities. Their combination provides the basis for future development paths for the agri-food sector.

Digital Integration

Digital integration allows uninterrupted information flow from the source to the point of use. Sensors could pick up a problem anywhere while the processing of data, the analysis of results and the determination of action to solve the problem could be done at any other place with no break in technology. In market activities, two of the most powerful applications are known as EDI (Electronic Document Interchange) and the integration of ERP (Enterprise Resource Planning) between business partners

EDI describes technologies for the automatic exchange of digital documents. While EDI in itself is not new, its technological standardization based on the internet technology and the communication language XML opens the way for a broad based acceptance and implementation. Initially, the primary focus of EDI was efficiency improvement. However, it is a necessary requirement for the intensification of information flows in quality and food safety management between enterprises in the food supply chain.

ERP-systems (Enterprise-Resource-Planning) are the backbone of data processing systems in enterprises. EDI might be used to organize communication between ERP-systems of different enterprises along the supply chain. However, internet technology has supported client-server applications which allow suppliers to directly link into the

ERP-systems of their customers (as if they were internal to the customer's business) and vis-a-versa. This allows the organization of highly integrated supply chains with improvements in efficiency, flexibility and communication between business partners. As an example, information on an enterprise's stocks are available to business partners at the same time they are available to the enterprise itself.

Multimedia Interaction

The integration of multimedia allows to combine documents, tables, graphs, sound, video and online multi-media communication capabilities as in videoconferencing etc. Many of today's advanced technologies that rely on keyboard input and character/graph-based output have not been able to live up to their promise because of missing multi-media capabilities. This is especially true in situations were IT applications are designed to directly support human communication needs. Examples include implementations of electronic trade platforms for food products, where trading traditionally is done over the phone, sometimes with different partners simultaneously. The switch from phone to keyboard/screen includes a loss in actual trading efficiency that needs to be matched by efficiency gains at other parts of the trading process before any overall gain in efficiency can be reached. Enterprises have discovered such problems earlier than promoters of new technologies and refused to accept present stage electronic trade platform technology for such products.

Electronic communication networks

Electronic communication networks are the infrastructure of the emerging information sphere that develops parallel to our visible world on the internet. They connect enterprises, knowledge bases, virtual meeting places, shops and any other unit that develops inside this sphere. However, the networks are not just for connection but they themselves together create a platform in its own right. This feature has not yet been utilized on a larger scale. Applications on the network still model (too) closely applications of the visible world with all its boundaries and limitations. A case in point is e-commerce technology with its common 'platform' approach that resembles a trading room.

Information portal technology

The information world successively develops parallel to the traditional physical world. In the internet, portals resemble entrance doors into the internet's knowledge sphere (Fritz et al., 2001). The organization of portals determines which parts of the sphere are directly accessible or indirectly by following pre-defined links. Portals may serve horizontal or vertical communication or information needs (Poignee et al., 2003)..

They might be open or limited to certain user groups. The set of different portals that serve the information needs of the sector and its individual enterprises determines the portal infrastructure of a sector and, in consequence, the sector's information situation. The design of an appropriate infrastructure is, therefore, a critical success factor for the utilization of the support potential of portal technology.

Virtual platforms for collaboration

The organization of virtual digital meeting places is a key feature of internet technology. They allow any type of collaborative activity among groups, including sharing of information, group discussions, project cooperation or document preparation. The e- commerce technology deals with meeting places for trading activities. It supports

(a) the establishment of virtual companies with joint sales or purchase activities ('shops') and new ways for cooperation and

(b) the organization of trading facilities ('electronic markets') where potential suppliers and customers meet for the trading of production inputs or food products.

The main difficulty is not the technical part but the organization of appropriate market rules (Shaw et al., 2000), the provision of appropriate trade information, the design of appropriate trade filters that determine the eligibility of participants and traded goods, and the organization of linkages between different electronic markets.

Agent technology

Agent technology assumes human tasks in information management. A specific application involves 'intelligent information agents' (Klusch, 2001). Intelligent information agents are software solutions, that search for information in distributed data sets or data bases and employ some level of autonomous and intelligent flexibility (Fritz et al., 2001). The search policy may be based on expert rules and resemble expert system technology. The value is in the automation of repetitious information search without human interaction. Potential applications deal with the focused collection of information from the information sphere of the internet and the information monitoring within food supply chains where information agents might search databases of suppliers or customers for information that might be relevant for the enterprise.

Organizational Integration

The various technological developments take place within a certain organizational market environment. The principal alternatives include 1. Food production within vertical food supply chains with a clearly defined and fixed group of participating enterprises (closed chains). 2. Food production within an open chain network with changing trading partners and dynamically evolving supply chain situations. The first alternative provides the best basis for an organized implementation of the technologies and to early reach a common high level of IT support. However, the second alternative is the dominant situation in the agri-food sector and, for various reasons, will remain so for the foreseeable future. This makes it not only more difficult to reach common agreements on investments in matching technology but requires more flexible solutions that (a) adapt to the changing conditions in the market environment, (b) allow for different implementation speeds, and (c) keep the need for common agreements as low as possible without slowing the development path for the more innovative sector participants.

The differences are much apparent in e-commerce and electronic trading activities. The concept of electronic trading platforms fits the trading environment of closed food supply chains but does not involve the flexibility that would best support the needs of open supply networks. However, combining the flexibility of the internet communication network with state of the art intelligent information agents allows the design of an e-commerce concept which could adequately match the dynamics of the food supply network (Hausen et al., 2001).

The core idea of this concept is to abandon the traditional e-commerce view of trading platforms as general meeting points but to view the internet network itself as the trading platform and to employ intelligent agents to establish temporarily evolving trading and communication links between electronic offices of participating enterprises. Intelligent agents search for trading opportunities, communicate trading interests and deliver responses to trading interests received from others through their agents. The system builds on individualized search and routing guides (programs) attached to each agent, which reflect the traders interaction interests with others on an individual case basis and on filters, which screen and filter incoming agents for source and type of information.

The concept of market places, where participants get together is replaced, in principle, by a virtual office concept where participants stay in their own virtual office environment and utilize communication and agent technology for communication with the virtual office environments of their trading partners. A supplier finds information on trading initiatives of his customers on his (the supplier's) individual virtual platform, a customer finds information on trading initiatives of his suppliers on his (the customer's) individual virtual platform and the agents assure the coordination. This concept heavily depends on appropriate network filters and agent communication guidance. Market participants have ownership of their platforms This facilitates the disassociation of individual platforms in case of sector problems.

The network is flexible in size and may steadily grow from a minimum base of a single company platform. This reduces the need for initial sector encompassing agreements. From a technological point of view, the concept integrates (and requires) elements from all of the development lines, except the multi-media element that would be an add-on for improved efficiency. It builds on digital integration, an electronic communication network, virtual offices as enterprise trading platforms, information agents for communication support, and portal technology for access to the network and the knowledge bases for trading support.

Implementation

The integration of a process-fitting combination of the technologies into business environments of agriculture and the food industry requires cooperation in various dimensions.

1. It requires cooperation between different areas of competence as, e.g., competence in information technology or competence in business and market management. The competence cooperation requirement is one of the main obstacles in the development and implementation of appropriate concepts for IT support in agri-food trade.

2. It requires cooperation between groups of enterprises from all stages of the food supply chain, extension organisations, market organisations, and other related services to make decisions regarding, among others, the organisation of communication processes in market activities, the communication content, the delivery technology (multi-media alternatives) or the portal infrastructure.

3. It requires cooperation between the providers of IT technologies and services to secure the technical feasibility and efficiency of digital integration and to adapt technology to content.

4. It requires cooperation among small-scale enterprises, especially farmers, to be able to utilize the emerging digital environments and to open opportunities for virtual cooperation.

5. It requires cooperation between users, research, extension and system design to arrive at applications which best fit the operational needs of users. 6. It requires cooperation in the development of system marketing strategies and the development and implementation of training opportunities.

The multitude of cooperation needs cannot be organized in a comprehensive way. However, they need to be promoted and encouraged to support the adoption of technologies and to gain from IT support within the sector as soon and as much as possible. As most of the developments have a distinguished sector dimension, their implementation cannot be decided by individual enterprises alone. This asks for the engagement of organisations with sector- wide acceptance to take an active promotional role and to actively initiate and coordinate necessary cooperation activities. This brings us back to the initial part of the paper where we stated that the ultimate evolution of IT support is '... a social choice. Society will ultimately choose a potential outcome depending on decisions made on investment, acceptance, adoption and rejection...'. It is at this point where decisions have to be made and where responsibilities of society would have to come in. The link between the sectors organizational-technological development and society's responsibilities derives from society's interest in the driving forces for the technological developments discussed in the paper, including a.o., the sector's competitive efficiency, the need to assure food quality, food safety, and consumer trust, and the interest in better environmental control. These responsibilities might be represented by sector-based organisations or policy institutions with sector-interest.

Recirculating Aquaculture Technology is Warding off Future Seafood Shortages

One sees it at local groceries and fresh markets. It's evident on the price list for fresh fish at many restaurants.

Seafood prices are on the rise, spurred by growing demand, a rapidly rising world population, and increasing pressures on already-depleted fish populations.

Virginia Tech researchers and Virginia Cooperative Extension specialists are helping meet that growing demand with recirculating aquaculture technology.

In short, they're growing fish at land-lubbing locales. By moving seafood production away from oceans and estuaries and into a controlled setting, recirculating aquaculture makes it possible to ward off future seafood shortages, safeguard the environment, and provide a spark for new enterprise.

Growing fish

Recirculating aquaculture refers to a method of growing fish at high densities under controlled conditions in indoor tanks. This approach to fish production minimizes the amount of water and land needed, and greatly expands the opportunities to grow fish in geographic areas that are normally unsuitable locations for seafood harvesting.

Progress in this field is essential. According to The Food and Agriculture Organization of the United Nations, the current pace of seafood consumption — 7 kilograms (15.4 pounds) per capita for the United States and 16 kilograms (35.2 pounds) worldwide — is not sustainable, and it necessitates the development of alternative ways to bring fish and shellfish into the market.

Virginia Tech is unique in the depth and breadth of recirculating aquaculture research and Extension it supplies.

Facilities in Detail

Blacksburg, Va.

One of the largest facilities dedicated to recirculating aquaculture research is the Virginia Tech Aquaculture Center (VTAC) in Blacksburg, Va. VTAC supports graduate-level research on a variety of fish species, including tilapia and cobia.

VTAC facilities can accommodate research in fresh, brackish, and marine waters under temperate, warm, and tropical conditions. Current research on cobia includes finding alternative protein and lipid sources for the fast-growing fish, as well as research on immunology and development.

The Aquatic Medicine Laboratory at the Virginia-Maryland Regional College of Veterinary Medicine (VMRCVM) in Blacksburg is a fully functional fee-for-service laboratory. It provides diagnostic services to aquaculturists, home aquarists, and researchers.

The laboratory specializes in the detection and evaluation of diseases affecting marine and freshwater foodfish species. It also works to identify parameters for the evaluation of fish health, including comparative blood chemistry and immunologic profiles for healthy and ailing fish.

Hampton, Va.

Located on the Hampton, Va., waterfront, the Virginia Seafood Agricultural Research and Extension Center (VSAREC) specializes in larval rearing of marine species, such as summer flounder and cobia.

VSAREC faculty members consistently achieve some of the highest survival rates documented during this fragile stage of the life cycle. The VSAREC not only provides fish for its own ongoing aquaculture research projects, but also supplies juvenile fish to the VTAC facility.

Additionally, the VSAREC hosts residential training programs for industry hatchery managers and graduate students, and conducts annual intensive marine larviculture production workshops.

Saltville, Va.

Situated in Virginia's mountainous west, the Southwest Virginia Aquaculture Research and Extension Center offers prospective growers concrete aquaculture production information.

This facility functions as a working freshwater recirculating aquaculture venture raising yellow perch. The center reaches out to area tobacco and dairy farmers and landowners interested in finding an alternative income-generating crop. They also offer hands-on learning opportunities and ongoing Extension support for new aquaculture businesses in the region.

The laboratory also investigates the potential usefulness of new drugs for use with cultured fish, develops vaccines for disease prevention, and explores the crossover potential of diseases such as mycobacteriosis, which can harm fish population and infect humans who encounter contaminated waters.

Industry Impact

Laboratory facilities, research programs, short courses, and diagnostic centers are geared to make a difference for the industry while adding to the knowledge base on animal physiology, management principles, and sustainable agriculture.

One way this knowledge is being put to work is at Blue Ridge Aquaculture facility in Martinsville, Va. The newest venture for Blue Ridge Aquaculture is a project to use tilapia effluents to support shrimp aquaculture.

Virginia Sea Grant and the U.S. Department of Agriculture will support the research at the shrimp culture facility that a multidisciplinary team of Virginia Tech researchers will conduct. They ultimately hope to establish the feasibility of shrimp/ tilapia co-culture on a commercial scale.

Through the support of these facilities, Virginia Tech researchers throughout Virginia will continue to make inroads into the science and application of recirculating aquaculture and to lead the way in disseminating this information to aquaculturists around the world.

Grain Cleaners

Summary of Harvey Gjesdal Five-in-one Rotary Seed Cleanerr

Quality of Work: The Gjesdal Five-In-One Rotary seed cleaner was effective in removing contaminants from spring wheat, oats and canola. Certified No. 2 grade was obtainable with acceptable losses in screenings. Certified No. 1 status could be obtained if problems such as ergot in wheat, bosom oats and wild oats in oats, and ball mustard in canola were not present or if other machines were used in conjunction with the Gjesdal machine.

Rate of Work: Capacity of the cleaner varies depending upon the crop, contaminants and grade desired, Best results with the samples tested were obtained with incoming feed rates of 44 bu/h (1.2 t/h) in wheat, 45 bu/h (0.7 t/h) in oats, and 10 bu/h (0.2 t/h) in canola. Ease of Operation and Adjustment: Adjustment of settings such as scalper drum speed and indent cylinder trough position were easily accomplished. However, changing grader shells or indent cylinders was awkward and time consuming.

Power Requirements: Two electric motors, 3/4 hp (0.56 kW) and 1 hp (0.75 kW) were used.

Operator's Manual: The manual supplied with the test machine provided a good explanation of how the machine functioned and some suggestions of possible settings for various crops. More information concerning service, safety and grader shell and indent cylinder replacement was needed.

Operator Safety: The Gjesdal cleaner was safe to operate if normal safety precautions were observed.

Mechanical History: No durability problems occurred during testing. However, the scalper drive belt required more tension and a very slight wobble developed in the indent cylinders near the end of the test.

Summary of Berglund Industrial Supply Company Combination Grain and Seed Cleaner

Quality of Work: The Bisco Combination Seed Cleaner was effective in removing contaminants from spring wheat, oats and canola. Certified No. 2 grade was obtainable with acceptable losses in screenings. Certified No. 1 status could be obtained if problems such as ergot in wheat, bosom oats and wild oats in oats, and ball mustard in canola were not present or if other machines were used in conjunction with the Bisco machine.

Rate of Work: Capacity of the cleaner varies depending upon the crop, contaminants and grade desired. Best results with the samples tested were obtained with incoming feed rates of 45 bu/h (1.2 t/h) in wheat, 50 bu/h (0.8 t/h) in oats, and 20 bu/h (0.4 t/h) in canola.

Ease of Operation and Adjustment: Adjustment of settings such as screen deck slope, air velocity and indent cylinder trough position were easily accomplished. Approximately 5 to 10 minutes were needed to change the screens.

Power Requirements: Two, 3/4 hp (0.56 kW) and one, 1 hp (0.75 kW) electric motors were used.

Operator's Manual: The manual supplied with the test machine provided an explanation of how the indent cylinder functioned and a parts list. More information concerning operation, adjustment, service and safety was needed.

Operator Safety: The Bisco cleaner was safe to operate if normal safety precautions were observed. Extra caution was needed when adjusting the controls of the aspirator on top of the machine.

Mechanical History: No durability problems occurred during testing.

Summary of Forever 42 Dustless Grain and Grass Seed Cleaner

Quality of Work: The Forever 42 Dustless Grain and Grass Seed Cleaner was effective in removing most contaminants from spring wheat, oats and canola. Canada No. 2 grade was easily obtainable with acceptable losses in screenings. Canada No. 1 status could be obtained if problems such as barley in wheat, wild oats and wheat in oats, and ball mustard in canola were not present.

Rate of Work: Capacity of the cleaner varies depending upon the crop, contaminants and grade desired. A Canada No. 2 grade was achieved with incoming feed rates of 60 bu/h (1.6 t/h) in wheat, 95 bu/h (1.4 t/h) in oats, and 40 bu/h (0.9 t/h) in canola.

Ease of Operation and Adjustment: Adjustment of the two fans was very fast and simple. The four screens took one person about 5 minutes to replace.

Power Requirements: One 5.0 hp (3.7 kW) electric motor was used to drive the cleaner and had adequate power.

Operator's Manual: The operator's instructions consisted of one page of parts description and suggested screens for various crops. Much more information on operation, adjustment, service and safety was required.

Operator Safety: The Forever cleaner was safe to operate if normal precautions were observed and the optional belt and pulley shields were properly installed.

Mechanical History: No durability problems occurred during testing.

Grain and Forage Dryers

Summary of Vertec VT 5600 R Grain Cleaner

Functional Performance: Functional performance of the Vertec VT 5600 R (1979 model) continuous flow grain dryer was very good in wheat, barley, rapeseed and HyBrid 3996 corn.

Drying Capacity: The rated drying capacity of the Vertec VT 5600 R was 9.1 t/h (334 bu/h) in wheat, 8.1 t/h (372 bu/h) in barley, 3.7 t/h (163 bu/h) in rapeseed and 5.3 t/h (209 bu/h) in corn. No grain damage occurred when operating at the manufacturer's recommended settings.

Fuel Consumption: Specific fuel consumption at rated drying capacity or the amount of propane required to dry a quantity of grain was 12.3 L/t (7.4 gal/100 bu) in wheat, 13.2 L/t (6.3 gal/100 bu) in barley, 10.0 L/t (5.0 gal/100 bu) in rapeseed and 24.7 L/t (13.8 gal/100 bu) in corn. This corresponded to a fuel consumption of 112 L/h (24.6 gal/h) in wheat, 107 L/h (23.5 gal/h) in barley, 37 L/h (8.1 gal/h) in rapeseed and 131 L/h (28.8 gal/h) in corn.

Energy Requirements: Specific energy consumption at rated capacity or the total energy required to remove a quantity of water from the grain, was 4700 kJ/kg (2000 Btu/lb) in wheat, 5100 kJ/kg (2200 Btu/lb) in barley, 4600 kJ/kg (2000 Btu/lb) in rapeseed and 4400 kJ/kg (1900 Btu/lb) in corn.

Ease of Operation and Adjustment: The Vertec VT 5600 R was easy to transport and set up. Burner performance was good and provided a steady and uniform grain drying temperature in most conditions. Optional equipment was required to dry corn at temperatures below 0° C (32°F). The grain drying temperature was easy to set, but had to be set slowly. Grain cooling was good and easy to adjust. Ease of filling and unloading was very good. Grain flow was uniform and adequate, but sensitive to adjust. Ease of cleaning the Vertec VT 5600 R was excellent. Lubrication points were accessible and easy to service.

Safety: The Vertec VT 5600 R was safe to operate as long as the manufacturer's safety instructions were followed. The sound level at the operator's station was 69 dBA.

Operator Manual: The operator's manual was well illustrated, clearly written and contained much useful information.

Mechanical History: The rear cross auger drive belt broke once during the test.

Summary of Superb S500C (1980 Model) Grain Dryer

Functional Performance: Functional performance of the Superb S500C (1980 model) continuous flow grain dryer was very good in wheat, barley, rapeseed and Hybrid 3996 corn.

The rated drying capacity of the Superb S500C was 9.4 t/h (345 bu/h) in wheat, 7.0 t/h (332 bu/h) in barley, 4.8 t/h (212 bu/h) in rapeseed and 4.8 t/h (189 bu/h) in corn. Grain damage occurred in wheat when operating at the manufacturer's setting.

Fuel Consumption: Specific fuel consumption at rated capacity, or the amount of propane required to dry a quantity of grain, was 8.2 L/t (4.9 gal/100 bu) in wheat, 8.4 L/t (4.0 gal/100 bu) in barley, 7.9 L/t (3.9 gal/100 bu) in rapeseed and 18.9 L/t (10.5gal/100 bu) in corn. This corresponded to a fuel consumption of 77 L/h (16.9 gal/h) in wheat 59 L/h (12.9 gal/h) in barley, 38 L/h (8.3 gal/h) in rapeseed and 91 L/h (19.9 gal/h) in corn.

Energy Requirements: Specific energy consumption at rated capacity, or the total energy required to remove a quantity of water from the grain was 3300 kJ/kg (1400 Btu/lb) in wheat, 3 400 kJ/kg (1500 Btu/lb) in barley, 3800 kJ/kg (1600 Btu/lb) in rapeseed and 3500 kJ/kg (1500 Btu/lb) in corn.

Ease of Operation and Adjustment: The Superb S500C was easy to install, but was not equipped for transporting, Burner performance was very good and provided a steady and uniform grain drying temperature in all conditions. Grain drying temperature was easy to set and was adequate for all conditions. Grain cooling was very good and easy to adjust. Ease of filling and unloading was good, Grain flow was uniform, adequate and easy to adjust. Ease of cleaning the Superb S500C was good. Small grains and fines partially plugged the grain chamber screens, and had to be cleaned out regularly. Lubrication points were accessible and easy to service.

Safety: The Superb S500C was safe to operate as long as the manufacturer's safety instructions were followed. The sound level at the operator's station was 85 dBA.

Operators Manual: The operator's manual was well illustrated, clearly written and contained much useful information.

Mechanical History: Three minor durability problems occurred during the tests.

Summary of Moridge 8440 Grain Dryer

Functional Performance: The Moridge 8440 recirculating batch grain dryer was very good in wheat, barley, rapeseed and Hybrid 3996 corn.

Drying Capacity: The rated drying capacity of the Moridge 8440 was 127 bu/h (3.5 t/h) in wheat, 134 bu/h (2.9 t/h) in barley, 124 bu/h (2.8 t/h) in rapeseed and 87 b/h (2.2 t/h) in corn.

Fuel Consumption: At rated drying capacity, the specific fuel consumption or the amount of propane required to dry a quantity of grain was 5.4 gal/100 bu (9.1 L/t) in wheat, 5.1 gal/100 bu (10.6 L/t) in barley, 4.2 gal/100 bu (8,4 L/t) in rapeseed and 11.4 gal/100 bu (20 4 L/t) in corn. This corresponds to a fuel consumption of 7.0 gal/h (32 L/ h) in wheat, 6.6 gal/h (30 L/h) in barley, 5.3 gal/h (24 L/h) in rapeseed and 9.9 gal/h (45 L/h) in corn.

Energy Consumption: At rated capacity, the specific energy consumption or the total energy required to remove a quantity of water from the grain, was 1500 Btu/lb (3600 kJ/kg) in wheat, 1800 Btu/lb (4200 kJ/kg) in barley, 1500 Btu/lb (3600 kJ/kg) in rapeseed and 1600 Btu/b (3700 kJ/kg) in corn.

Grain Quality: A grade loss occurred in commercial red spring wheat when operating at the manufacturers recommended temperature settings. At temperature settings that did not reduce the grade of commercial red spring wheat the rated drying capacity decreased 30 to 40% while specific fuel and energy consumption increased by 10 to 20%.

Ease of Operation and Adjustment: The Moridge 8440 was easy to transport and setup. Burner performance was very good and provided a steady and uniform drying air temperature in most conditions. Adequate drying air temperatures could not be obtained when drying corn at outside temperatures below 20 F (-7oC). The drying air temperature was easy to set. Grain cooling was very good. Ease of filling and unloading was good. Grain recirculation was adequate for most conditions. Ease of cleaning the Moridge 8440 was fair. Lubrication points were accessible and fairly easy to service.

Safety: The Moridge 8440 was safe to operate as long as the manufacturer's safety instructions were followed. The sound level at the operator's station was 102 dBA. It is recommended that an operator wear ear protection when working near the Moridge 8440.

Operator Manual: The operator manual was well illustrated, clearly written and contained much useful information.

Mechanical History: Seven minor mechanical problems occurred during the tests.

Summary of GT 380 Grain Dryer

Functional Performance: The GT 380 recirculating batch grain dryer was very good in wheat, barley, rapeseed and Hybrid 3996 corn.

Drying Capacity: The rated drying capacity of the GT 380 was 99 bu/h (2.7 t/h) in wheat, 124 bu/h (2.7 t/h) in barley, 110 bu/h (2.5 t/h) in rapeseed and 75 bu/h (1.9 t/h) in corn.

Fuel Consumption: At rated drying capacity, the specific fuel consumption or the amount of propane required to dry a quantity of grain was 6.4 gal/100 bu (10.6 L/t) in wheat. 4.9 gal/100 bu (10.3 L/t) in barley, 4.7 gal/100 bu (9.5 L/t) in rapeseed and 12.7gal/100 bu (22.7 L/t) in corn. This corresponds to a fuel consumption of 6.4 gal/h (29 L/h) in wheat, 6.2 gal/h (28 L/h) in barley, 5.3 gal/h (24 L/h) in rapeseed and 9.5 gal/h (43 L/h) in corn.

Energy Consumption: At rated capacity, the specific energy consumption or the total energy required to remove a quantity of water from the grain, was 1900 Btu/lb (4300 kJ/kg) in wheat, 1800 Btu/lb (4100 kJ/kg) in barley, 1800 Btu/lb (4100 kJ/kg) in rapeseed and 1800 Btu/lb (4100 kJ/kg) in corn.

Grain Quality: No grade loss occurred when operating at the manufacturer's recommended temperature settings.

Ease of Operation and Adjustment: The GT 380 was difficult to transport, but easy to set up. Burner performance was very good and provided a steady and uniform drying air temperature in most conditions. Adequate drying air temperatures could not be obtained when drying corn at outside temperatures below 20°F (-7°C). The drying air temperature was easy to set. Grain cooling was very good. Ease of filling and unloading was good. Grain recirculation was adequate for all conditions. Ease of cleaning the GT 380 was good. Lubrication points were accessible and easy to service.

Safety: The GT 380 was safe to operate as long as the manufacturer's safety instructions were followed. The sound level at the operator's station was 103 dBA. It is recommended that an operator wear ear protection when working near the GT 380.

Operator Manual: The operator manual was well illustrated, clearly written and contained much useful information.

Mechanical History: Six minor mechanical problems occurred during the tests.

Summary of Adams Grain Dryer

Drying Capacity: The rated drying capacity of the Adams Grain Dryer was 82 bu/h (2.2 t/h) in wheat, 98 bu/h (2.1 t/h) in barley, 82 bu/h (1.9 t/h) in rapeseed and 61 bu/h (1.6 t/h) in Hybrid 3998 corn.

Fuel Consumption: At the rated drying capacity, the specific fuel consumption or the amount of propane required to dry a quantity of grain was 7.0 gal/100 bu (11.7 L/t) in wheat, 5.5 gal/100 bu (11.5 L/t) in barley, 5.5 gal/100 bu (11.0 L/t) in rapeseed and 13.3 gal/100 bu (23.8 L/t) in corn. This corresponds to a fuel consumption of 5.7 gal/h (26 L/h) in wheat, 5.4 gal/h (25 L/h) in barley, 4.5 gal/h (20 L/h) in rapeseed and 8.1 gal/h (37 L/h) in corn.

Energy Consumption: At rated drying capacity, the specific energy consumption or the total energy required to remove a quantity of water from the grain, was 2000 Btu/lb (4700 kJ/kg) in wheat, barley and rapeseed and 1900 Btu/lb (4400 kJ/kg) in corn.

Quality of Work: No grade loss occurred in the grains tested when operating at the manufacturer's recommended drying air temperature settings. The drying air temperature was very uniform and adequate for all drying conditions encountered, but was slightly lower than the temperature setting.

Ease of Operation and Adjustment: Ease of assembly and installation was very good. The Adams Grain Dryer was mounted on skids for transporting short distances. Ease of filling and discharge was fair. Ease of drying was fair as regular supervision was required. The drying air temperature was difficult to set. Ease of cooling was good. Ease of cleaning the Adams Grain Dryer was good as the air plenum and auger sump required cleaning weekly. Only seasonal lubrication was required making the ease of servicing excellent.

Power Requirements: The Adams Grain Dryer required 10.4 hp (7.8 kW) when operated on 230 V AC electrical power.

Safety: The Adams Grain Dryer was safe to operate as long as the manufacturer's limited safety instructions and common safety practices were followed. The sound level at the operator's station was 107 dBA. It is recommended that the operator wear ear protection when working near the Adams Grain Dryer.

Operator Manual: The operator manual was incomplete and poorly written, providing limited useful information on operation, adjustments and safety.

Mechanical History: Seven mechanical problems occurred during the test.

Summary of Drymor Redbird Grain Dryer

Drying Capacity: The rated drying capacity of the Drymor Redbird was 137 bu/h (3.7 t/h) in wheat, 153 bu/h (3.3 t/h) in barley, 101 bu/h (2.3 t/h) in rapeseed and 104 bu/h (2.6 t/h) in Hybrid 3996 corn.

Fuel Consumption: At rated drying capacity, the specific fuel consumption or the amount of propane required to dry a quantity of grain was 7.4 gal/100 bu (12.4 L/t) in wheat, 5.8 gal/100 bu (12.1 L/t) in barley, 6.3 gal/100 bu (12.6 L/t) in rapeseed and 13.0 gal/100 bu (23.3 L/t) in corn. This corresponds to a fuel consumption of 10.1 gal/h (46 L/h) in wheat, 8.8 gal/h (40 L/h) in barley, 6.4 gal/h (29 L/h) in rapeseed and 13.7 gal/h (62 L/h) in corn.

Energy Consumption: At rated drying capacity, the specific energy consumption or the total energy required to remove a quantity of water from the grain, was 2100 Btu/lb (4900 kJ/kg) in wheat, 2300 Btu/lb (5400 kJ/kg) in barley, 2000 Btu/lb (4700 kJ/kg) in rapeseed and 1800 Btu/lb (4200 kJ/kg) in corn.

Quality of Work: No grade loss occurred when drying commercial rapeseed or feed barley and corn. A grade loss did occur in commercial red spring wheat when operating at the manufacturer's recommended drying air temperature setting. At a drying air temperature that did not reduce the grade of commercial red spring wheat, the rated drying capacity would decrease 40 to 50%, while specific fuel and energy consumption would increase by 10 to 20%. The drying air temperature was uniform and adequate for all drying conditions encountered, but was slightly lower than the gauge reading.

Ease of Operation and Adjustment: Ease of assembly and installation was very good. The Drymor Redbird was not equipped for transporting. The automatic controls made the ease of filling, drying, cooling and discharge very good. Supervision was only required on the first run each time grain conditions changed to determine the new control settings. The drying air temperature was easy to set. Ease of cleaning the Drymor was fair and it required weekly cleanup. Lubrication points were accessible and daily servicing took 2 minutes, making the ease of servicing very good.

Power Requirements: The Drymor Redbird required 13.7 hp (10.2 kW) when operated on 230 V AC electrical power.

Safety: The Drymor Redbird was safe to operate as long as the manufacturer's safety instructions were followed. The dryer was quiet to work around. The operator's station was located near the fan resulting in a sound level of 90dBa.

Operator Manual: The operator manual was well illustrated, clearly written and contained much useful information.

Mechanical History: One minor mechanical problem occurred during the test.

Summary of Super-B AS-600G Grain Dryer

Drying Capacity: The rated drying capacity of the Super-B AS-600G was 143 bu/h (3.9 t/h) in wheat, 179 bu/h (3.9 t/h) in barley, 140 bu/h (3.2 t/h) in rapeseed, and 126 bu/h (3.2 t/h) in Hybrid 2120 corn.

Fuel Consumption: At rated drying capacity, the specific fuel consumption or the amount of propane required to dry a quantity of grain was 6.4 gal/100 bu (10.7 L/t) in wheat, 5.8 gal/100 bu (12.1 L/t) in barley, 4.6 gal/100 bu (9.3 L/t) in rapeseed, and 11.0 gal/100 bu (19.6 L/t) in corn. This corresponds to a fuel consumption of 9.2 gal/h (42 L/h) in wheat, 10.3 gal/h (47 L/h) in barley, 6.6 gal/h (30 L/h) in rapeseed, and 13.6 gal/h (62 L/h) in corn.

Energy Consumption: At rated drying capacity, the specific energy consumption or the total energy required to remove a quantity of water from the grain was 1800 Btu/lb (4200 kJ/kg) in wheat, 2000 Btu/lb (4700 kJ/kg) in barley, 1700 Btu/lb (3900 kJ/kg) in rapeseed, and 1500 Btu/lb (3600 kJ/kg) in corn.

Quality of Work: No grade loss occurred in the grains tested when operating at the manufacturer's recommended drying air temperature settings. The drying air temperature was very uniform and adequate for all drying conditions encountered and was close to the temperature setting.

Ease of Operation and Adjustment: Ease of assembly and installation was excellent. The Super-B AS-600G was not equipped for transporting. The automatic controls made the ease of filling, drying, cooling, and discharge very good. Supervision was only required on the first run each time grain conditions changed to determine the new control settings. The drying air temperature was easy to set. Ease of cleaning the Super-B was very good. However, the screens required daily cleaning in rapeseed. Lubrication points were accessible and semi-monthly servicing took two minutes, making the ease of servicing very good.

Power Requirements: The Super-B AS-600G required 15.8 hp (11.8 kW) when operated on 230 V AC electrical power.

Safety: The Super-B AS-600G was safe to operate as long as the manufacturer's safety instructions were followed. The sound level at the operator's station was 102 dBA. It is recommended that the operator wear ear protection when working near the Super-B AS-600G.

Operator's Manual: The operator's manual was good. It was sufficiently illustrated, clearly written, and contained much useful information. However, it contained a few errors and omissions.

Mechanical History: Two minor mechanical problems occurred during the test.

Summary of Bader Bin

Drying uniformity was very good. System pressure loss was 0.5 to 1.5 in wg (125 to 375 Pa) at an airflow of 2000 cfm (940 L/s) in wheat. The natural drying system could be factory installed only in Bader bins and could not be field installed. Ease of cleaning the natural air drying system was very good. No operator's manual was provided and no mechanical problems were encountered. The bin and air distribution system were epoxy coated for storing fertilizer.

Summary of Grain Guard GG 7000 Bin

Drying uniformity was very good. System pressure loss was 0.5 to 1.5 in wg (125 to 375 Pa) at an airflow of 2000 cfm (940 L/s) in wheat. The natural air drying system was factory installed by the bin manufacturer and could be field installed in some bins.

Ease of cleaning the natural air drying system was good. No operator's manual was provided and no mechanical problems were encountered. The system could be epoxy coated as an option for storing fertilizer.

Summary of Keho Diamond Bin

Drying uniformity was good. System pressure loss was 0.5 to 1.5 in wg (125 to 375 Pa) at an airflow of 2000 cfm (940 L/s) in wheat. The natural air drying system was field installed by two men in about 4 hours. It could be installed in any bin.

Ease of cleaning the natural air drying system was very good. The operator's manual was very good, although installation instructions were not clear. No mechanical problems were encountered. The system had to be removed for storing fertilizer.

Summary of Lode-King High Flow Bin

Drying uniformity was very good. System pressure loss was 0.5 to 1.5 in wg (125 to 375 Pa) at an airflow of 2000 cfm (940 L/s) in wheat. The natural air drying system was factory installed and could be field installed in some bins.

Ease of cleaning the natural air drying system was very good. No operator's manual was provided and no mechanical problems were encountered. The system had to be removed for storing fertilizer.

Summary of Trail Rite Mark IV Bin

Drying uniformity was good. System pressure loss was 1.0 to 2.0 in wg (250 to 5005 Pa) at an airflow of 2000 cfm (940 L/s) in wheat. The distribution duct was factory installed and could be field installed in some bins.

Ease of cleaning the natural air drying system was very good. The short slide-grate crank handle was potentially unsafe to operate when unloading. No operator's manual was provided and no mechanical problems were encountered. The system had to be removed for storing fertilizer.

Summary of Weninger Horizontal

Drying uniformity was good. System pressure loss was 0.5 to 1.5 in wg (125 to 375 Pa) at an airflow of 2000 cfm (940 L/s) in wheat. Installing the natural air drying system took two men about 2 hours, it could be field installed in some bins.

Ease of cleaning the natural air drying system was good. Instructions were good, but did not include complete natural air drying guidelines, No mechanical problems were encountered. The system had to be removed for storing fertilizer.

Summary of Weninger Rocket Bin

Drying uniformity was very good. System pressure loss was 0.5 to 1.4 in wg (125 to 375 Pa) at an airflow of 2000 cfm (940 L/s) in wheat. The natural air drying system was factory installed but could be field installed in Weninger bins. Ease of cleaning the natural air drying system was good. Instructions were very good, but did not include complete natural air drying guidelines. No mechanical problems were encountered. The system had to be removed for storing fertilizer.

Summary of Keho Cyclone Bin

Average-drying time was 0.95 times the reference bin. Thorough-drying time was greater than 1.75 times the reference bin.

Airflow rate was 1.05 cfm/bu (13.7 L/s m^3) with a typical 3 hp (2.2 kW) fan. Drying uniformity results indicated that parts of the top layer of grain took grater than 35% longer to dry than if airflow uniformity was ideal.

The natural air drying system could be field installed in any bin. Ease of installing the fan was fair. Ease of cleaning was very good.

Installation instructions were provided and no mechanical problems were encountered. The system had to be removed for storing fertilizer.

Summary of Grain Guard GG 7000 Bin

Average-drying time was 0.75 times the reference bin. Thorough-drying time was 1.05 times the reference bin. Airflow rate was 1.20 cfm/bu (15.6 L/s m^3) with a typical 3 hp (2.2 kW) fan. Drying uniformity results indicated that parts of the top layer of grain took greater than 35% longer to dry than if airflow uniformity was ideal.

The natural air drying system could be field installed in some bins. Ease of installing the fan was good. Ease of cleaning was very good.

No operator's manual was provided and no mechanical problems were encountered. The system could be epoxy coated for storing fertilizer.

Summary of Naicam Bin

Average-drying time was 0.95 times the reference bin. Thorough-drying time was 1.10 times the reference bin.

Airflow rate was 1.05 cfm/bu (13.7 L/s m^3) with a typical 3 hp (2.2 kW) fan. Drying uniformity results indicated that parts of the top layer of grain took greater than 35% longer to dry than if airflow uniformity was ideal.

The natural air drying system could be field installed in some bins. Ease of installing the fan was fair. The system had to be removed for storing fertilizer.

Summary of Taylor Bin

Average-drying time was 0.90 times the reference bin. Thorough-drying time was 0.80 times the reference bin.

Airflow rate was 1.15 cfm/bu (15.0 L/s m^3) with a typical 3 hp (2.2 kW) fan. Drying uniformity results indicated that parts of the top layer of grain took 14% longer to dry than if airflow uniformity was ideal.

The natural air drying system could be field installed in any bin. Ease of installing the fan was poor. Ease of cleaning was very good. No operator's manual was provided and no mechanical problems were encountered. The system could be epoxy coated for storing fertilizer.

Summary of Univision Ultra-Dry (Series 2000) Bin

Average-drying time was 0.70 times the reference bin. Thorough-drying time was 0.75 times the reference bin.

Airflow rate was 1.40 cfm/bu (18.2 L/s m^3) with a typical 3 hp (2.2 kW) fan. Drying uniformity results indicated that parts of the outside layer of rain took greater than 35% longer to dry than if airflow uniformity was ideal.

The natural air drying system could be factory installed in Univision bins, but could not be field installed. Ease of installing the fan was good. Ease of cleaning was very good.

No operator's manual was provided and no mechanical problems were encountered. The system could be epoxy coated for storing fertilizer.

Summary of Bader Hooded Bin

Average-drying time was 0.70 times the reference bin. Thorough-drying time was 1.20 times the reference bin.

Airflow rate was 1.15 cfm/bu (15.0 L/s m^3) with a typical 3 hp (2.2 kW) fan. Drying uniformity results indicated that parts of the top layer of grain took 32% longer to dry than if airflow uniformity was ideal.

The natural air drying system could be field installed in Bader bins. Ease of installing the fan was fair. Ease of cleaning was very good.

No operator's manual was provided and no mechanical problems were encountered. The system could be epoxy coated for storing fertilizer.

Summary of Grehl Bin

Average-drying time was 1.10 times the reference bin. Thorough-drying time was 1.20 times the reference bin.

Airflow rate was 0.90 cfm/bu (10.4 L/s m^3) with a typical 3 hp (2.2 kW) fan. Drying uniformity results indicated that parts of the top layer of grain took 6% longer to dry than if airflow uniformity was ideal.

The natural air drying system could be field installed in any bin. Ease of installing the fan/transition unit was fair. Ease of cleaning was poor.

No operator's manual was provided and no mechanical problems were encountered. The system could not be epoxy coated for storing fertilizer.

Summary of Harvest Hopper Bin

Average-drying time was 1.00 times the reference bin. Thorough-drying time was 1.30 times the reference bin.

Airflow rate was 0.85 cfm/bu (11.1 L/s m^3) with a typical 3 hp (2.2 kW) fan. Drying uniformity tests indicated that parts of the top layer of grain took 35% longer to dry than if airflow uniformity was ideal.

The natural air drying system could be factory installed in Harvest Hopper bins, but could not be field installed. Ease of installing the fan was very good. Ease of cleaning was very good.

No operator's manual was provided and no mechanical problems were encountered. The system could be epoxy coated for storing fertilizer.

Summary of Trail Rite V Aeration Bin

Average-drying time was 1.20 times the reference bin. Thorough-drying time was 1.10 times the reference bin.

Airflow rate was 1.00 cfm/bu (13.0 L/s m^3) with a typical 3 hp (2.2 kW) fan. Drying uniformity results indicated that parts of the top layer of grain took 16% longer to dry than if airflow uniformity was ideal.

The natural air drying system could be field installed in some bins. Ease of installing the fan was good. Ease of cleaning was very good.

No operator's manual was provided and no mechanical problems were encountered. The system could be epoxy coated for storing fertilizer.

Summary of Univision In-Line Bin

Average-drying time was 0.90 times the reference bin. Thorough-drying time was 0.90 times the reference bin.

Airflow rate was 1.10 cfm/bu (14.3 L/s m^3) with a typical 3 hp (2.2 kW) fan. Drying uniformity results indicated that parts of the top layer of grain took 17% longer to dry than if airflow uniformity was ideal.

The natural air drying system could be field installed in some bins. Ease of installing the fan was fair. Ease of cleaning was very good.

No operator's manual was provided and no mechanical problems were encountered. The system could be removed for storing fertilizer.

Summary of Univision Ultra-Dry Bin

Average-drying time was 0.70 times the reference bin. Thorough-drying time was 0.90 times the reference bin.

Airflow rate was 1.15 cfm/bu (15.0 L/s m^3) with a typical 3 hp (2.2 kW) fan. Drying uniformity results indicated that parts of the top layer of grain took 33% longer to dry than if airflow uniformity was ideal.

The natural air drying system could be field installed in Univision bins, but could not be field installed. Ease of installing the fan was good. Ease of cleaning was good. Some grain remained in the bin when the initial grain moisture content was above 20%.

Summary of Westeel Boot Bin

Average-drying time was 1.00 times the reference bin. Thorough-drying time was 1.20 times the reference bin.

Airflow rate was 0.85 cfm/bu (11.1 L/s m^3) with a typical 3 hp (2.2 kW) fan. Drying uniformity results indicated that parts of the top layer of grain took 26% longer to dry than if airflow uniformity was ideal.

The natural air drying system could be field installed in bins with a Westeel Boot. Ease of installing the fan was very good. Ease of cleaning was very good.

No operator's manual was provided and no mechanical problems were encountered.No operator's manual was provided and no mechanical problems were encountered. The system could be removed for storing fertilizer.

Chemical Fertilizer Applicators

Summary of Melroe 2710 Fertilizer Attachment

Quality of Work: Overall functional performance of the Melroe 2710 fertilizer attachment was fair. Its performance was reduced by large variations in application rate with changes in field slope and fairly high variation in distribution across the seeding width.

Application uniformity was not affected by field roughness, ground speed or level in the fertilizer box. Application rate was significantly affected by field slope and increased 53% when seeding up a 15° slope. The variation in application across the seeding width was just within acceptable limits.

Actual application rates were about 30 kg/ha (27 lb/ac) higher than those indicated in the manufacturer's calibration tables for all application rates when applying fertilizer types commonly used in the prairies. A large range of fertilizing rates were available with an adequate number of settings in the commonly used range.

Ease of Installation: Installation of the attachment on a drill was easy with adequate instructions supplied by the manufacturer.

Ease of Operation and Adjustment: The fertilizer box was very easy to clean. The feed baffles and dropout bottom were easily removed. The box was well sealed with a slight amount of moisture entering only in heavy rains.

Operators Manual: The operator's manual was very good. It contained comprehensive instructions on installation, adjustment, repair and maintenance.

Mechanical History: Only one minor mechanical problem occurred during functional evaluation.

Summary of Spierco M80 Fertilizer Attachment

Quality of Work: Overall functional performance of the Spierco M80 fertilizer attachment was very good. Performance was reduced by variation in application rates with changes in forward speed and by nonuniformity of distribution across the seeding width at low fertilizing rates.

Application rate was not affected by field roughness or level in the fertilizer box, and was only slightly affected by field slope. Application across the seeding width was uniform at high fertilizing rates but was non-uniform at low rates. Application uniformity was affected by side slope.

Actual application rates were slightly lower than indicated by the manufacturer's calibration. The calibration of the two boxes was slightly different. A large range of fertilizing rates was available with an adequate number of settings.

Ease of Operation and Adjustment: The fertilizer box was easy to clean, since the Screens and baffles were easily removed. The box was well sealed.

Operators Manual: The operator's manual contained comprehensive instructions on operation and maintenance.

Mechanical History: Only one minor mechanical problem with a drive cable occurred during the test.

Summary of Gandy 44-NDK 59 Fertilizer Attachment

Quality of Work: Overall functional performance of the Gandy 44-NDK 59 fertilizer attachment was very good. Performance was reduced by variation in application rates with changes in forward speed and by plugging of the delivery tubes when travelling down slopes greater than 10°.

Application rate was not affected by field roughness or level in the fertilizer box, and was only slightly affected by field slope unless delivery tubes plugged. Application across the seeding width was very uniform.

The manufacturer's calibration was accurate and a large range of fertilizing rates was available with an adequate number of settings.

Ease of Operation and Adjustment: The fertilizer box was easy to clean since the rotor bar and box bottom were easily removed. The box was well sealed.

Operators Manual: The operator's manual contained comprehensive instructions on operation and maintenance.

Mechanical History: No mechanical problems occurred during the test.

Summary of Fertilizer Banding Knife Attachment

Conventional Cultivator Chisel Point: Cultivator chisel points left an uneven field surface. Fertilizer was placed in bands 1.25 in (32 mm) wide and varied 0.40 in (10 mm) in depth when the chisel points were new. Installation on the cultivator was convenient. Fertilizer band width increased when the points were worn. Penetration was very good. Power requirements were usually less than for most banding knives.

Culti-Band Q0115: Fertilizer was placed in bands 0.75 in (19 mm) wide and varied 0.40 in (10 mm)in depth. Penetration was very good when the knife was new and fair after the knife became worn. Installation on the cultivator was convenient. Power requirements for new knives were about 20% greater than those for new chisel points. Power requirements for worn knives were about 6% less than for worn chisel points. Several replaceable knife tips were lost during the test.

Dutch Granular Knife Number 7: Fertilizer was placed in bands 1.0 in (25 mm) wide and varied 0.40 in (10 mm) in depth. Penetration was very good when the knife was new and fair after the knife became worn. Installation on the cultivator was convenient. Power requirements for new knives were about 15% greater than for new chisel points. Power requirements for worn knives were about 9% greater than for worn chisel points.

Gromor Deep Banding Knife: Fertilizer was placed in band 0.75 in (19 mm) wide and varied 0.40 in (10 mm) in depth. Penetration was very good when the knife was new and poor after the knife became worn. Initial installation on the cultivator was inconvenient. The cultivator shanks had to be cut off and modified to accommodate the Gromor deep banding knife. Power requirements for new knives were about 19% greater than for new chisel points. Power requirements for worn knives were about 16% greater than for worn chisel points.

Digger Bandit Snout 3: Fertilizer was placed in bands 1.0 in (2.5 mm) wide and varied 0.40 in (10 mm) in depth. Penetration was good when the knives were new and fair after the knives became worn. Installation on the cultivator was easy. Fertilizer feed tubes were not supplied. Power requirements for new knives were about 40% greater than for new chisel points. Power requirements for worn knives were about 70% greater than for worn chisel points.

JAD Opener: Fertilizer was placed in bands 1.0 to 1.25 in (25 to 32 mm) wide and varied 0.35 to 0.50 in (9 to 13 mm) in depth. Band width and thickness varied due to variations in knife manufacturing uniformity. Penetration was good when the knives were new and poor after the knives became worn. Installation on the cultivator was inconvenient due to limited access to the lower bolt hole. Power requirements for new knives were about 8% greater than for new chisel points, Power requirements for worn knives were about 3% higher than for worn chisel points. A weld on one knife support bracket failed when an obstruction was encountered.

TSL Deep Banding Knife: Fertilizer was placed in bands 0.75 in (19 mm) wide and varied 0.40 in (10 mm) in depth. Penetration was very good when the knives were new and poor after the knives became worn. Installation on the cultivator was easy, Power requirements for new knives were about 24% greater than for new chisel points. Power requirements for worn knives were about 32% greater than for worn chisel points.

Haley Deep Banding Knife: Fertilizer was placed in bands 0.75 in (19 mm) wide and varied 0.40 in (10 mm) in depth. Penetration was poor when the knives were new and poor after the knives became worn. Installation on the cultivator was inconvenient due to limited access to the support bracket bolt holes. Power requirements for new knives were about 20% greater than for new chisel points. Power requirements for worn knives were about 18% greater than for worn chisel points. Wear of the knife tips was excessive and caused plugging of the fertilizer tubes.

Summary of Valmar Model 4400 Fertilizer Attachment

Quality of Work: The accuracy of the fertilizer metering system was excellent. Fertilizer was applied at rates from 35 to 436 lb/ac (39 to 488 kg/ha) depending on density. The manufacturer's calibration chart was accurate. A convenient scale was supplied for determining density, and checking application rates in the field. Distribution across the width of the Valmar 4400 was very uniform with coefficient of variation's (CV) less than 5%.

Performance of the fertilizer delivery system was very good: The smooth plastic hoses did not plug, and did not damage the fertilizer.

Fertilizer placement depended upon the cultivator and banding openers used. Fan speed did not affect fertilizer placement.

Ease of Installation: Ease of installing the Valmar 4400 was very good. It took two men about 16 hours. It was easily adapted to the test cultivator. A loader or crane was required. Installation instructions were adequate.

Ease of Operation and Adjustment: Ease of filling the tank was good. The 9.2 ft (2.8 m) filling height was too high for most drill fills, but worked well for auger filling. The tank lid was weathertight. The tank access ladder was too short.

Ease of cleaning was very good. The metering rollers were easily removed, once the tank was empty. The venturis and air manifold were easy to clean. Some fertilizer caked on the metering rollers and venturis in cool, damp weather.

Monitoring was good. A fan tachometer, a low bin level alarm, and meter shutoffs were provided on the control console in the tractor. The meters and venturis could be viewed from the tractor, but the plexiglass shield became dirty and could not be kept clean. The high tank obstructed visibility of the cultivator frame.

Ease of transporting was very good. The Valmar 4400 did not interfere with normal transporting of the cultivator. The tank should be emptied before transporting to avoid excessive cultivator frame stress or tire overloading,

Application rate adjustment was very good. Fan adjustment was fair. With the pto drive, the full range of fan speeds could hp (7.5 kW) at a fan speed of 5000 rpm and tractor pto speed of 1100 rpm.

Operator Safety: Drive components were adequately shielded. Warning decals were supplied. The tank access ladder was too short, making it hazardous when filling the tank.

Operator's Manual: The operator's manual was very good. It was thorough, complete, and well illustrated.

Mechanical History: A few minor mechanical problems occurred during 70 hours of field work.

Summary of ACCU Lil Bander Fertilizer Attachment

Quality of Work: The accuracy of the fertilizer metering system was very good. Fertilizer was applied at rates from 24 to 345 lb/ac (27 to 385 kg/ha) depending on type. The calibration chart was accurate but could not be used for custom blended fertilizers. A method for field checking rates was not provided. Distribution across the width of the Accu Lil Bander was uniform with CV's less than 10%.

Performance of the fertilizer delivery system was very good. The smooth-lined plastic hoses did not plug, and the box had adequate height for gravity flow.

Fertilizer placement depended upon the cultivator and banding openers used.

Ease of Installation: Ease of installing the Accu Lil Bander was very good, it took one man about 8 hours to install each box. It was easily adapted to the test cultivator. A loader or crane was required. Written instructions and several sketches were adequate.

Ease of Operation and Adjustment: Ease of filling the tank was good. The 7.8 ft (2.4 m) filling height was too high for some drill fills, but worked well for auger filling. The large filler opening was weathertight, Ease of cleaning was fair. Fertilizer had to be swept or vacuumed out of the flat bottom. The metering wheels were easily removed.

Monitoring was good. Material flow could be viewed from the tractor. A sight glass helped to judge the amount of fertilizer in the box. The boxes obstructed visibility behind the cultivator.

Ease of transporting was very good. The boxes were offset to allow the cultivator wings to fold up normally. The wing lifts had to be slightly adjusted. The boxes should be emptied before transporting to avoid excessive cultivator frame stress or tire overloading.

Application rate adjustment was good. The multiple speeds were easy to set. Ease of maintenance was good. Daily servicing took about 3 minutes. No servicing information was supplied.

Power Requirements: The Accu Lil Bander had no measurable effect on the cultivator power requirements.

Operator Safety: No serious safety hazards were apparent with the machine. Climbing onto the cultivator for filling was hazardous.

Operator's Manual: The operator's manual was good. It contained information on installation, operation, adjustment, and troubleshooting. Safety information was not provided.

Mechanical History: A few minor problems occurred.

Summary of Clarke Dual Delivery System

Quality of Work: Penetration of the Clarke Dual Delivery System was good. Penetration depended entirely on the shank trip characteristics. A shank with a trip force of at least 750 lb (3340 N) is recommended.

Seed placement of the Dual Delivery System was very good. The seeding blade split the seed evenly into two rows 6 in (152 mm) apart. Bandwidth of the rows was slightly wider than that of a hoe drill. Use of the seed blades left ridges. This required the use of a harrow packer drawbar as a post seeding operation.

Fertilizer placement of the Dual Delivery System was very good. The fertilizer was placed directly between, and 2 in (51 mm) below the paired rows of seed.

Trash clearance was good. The use of the Clarke system increased the trash clearance capability of the test cultivator in normal conditions. Performance, though, was reduced in adverse wet or heavy trash conditions.

Operation of the Clarke system in stony conditions was very good. No major damage occurred to the deep banding blade or seed blade.

Ease of Operation and Adjustment: Ease of transporting the test cultivator with the Clarke Dual Delivery Systems mounted was poor. The deep banding blades had to be removed or secured up to allow for ground clearance while transporting.

The openings on the seeding blades did not plug during the test. The openings on the deep banding blades started to plug at the end of the test because of blade wear and operation in wet soils.

Power Requirements: Draft (drawbar pull) requirements depended on depth, field preparation, ground speed, soil type and moisture content. The draft on the Clarke system was 50 to 79% greater than a 16 in (406 mm) chisel sweep at the same seeding depth. Maximum tractor power requirements in primary conditions for one Clarke Dual Delivery System ranged from 7.1 to 9.4 hp (5.3 to 7.1 kW).

Ease of Installation: Ease of installing the Clarke Dual Delivery System was good. Two major modifications had to be made to the test cultivator to accommodate the Clarke Dual Delivery System.

The Clarke Dual Delivery System presented no safety hazard if normal safety precautions were observed.

Operators Manual: No operator's manual was provided.

Mechanical History: The deep banding knives were still useable after 25 acres (10 ha) per blade.

Summary of Morris Air Hoe Packer Attachment

Quality of Work: Penetration of the Morris Air Hoe Packer Attachment was good. Penetration of the hoe opener in primary conditions depended on the depth of the knife opener and the packing force. Secondary soil conditions improved penetration..

Seed placement of the air hoe packer attachment was good. With the wide sweep, the seed was split evenly into two rows 5 in (127 mm) apart. The rows were distinct with the bandwidth a little wider than that of a hoe drill. With the narrow sweep, the seed was placed in a row with an average bandwidth of 4 in (102 mm). The two rows were not distinct..

The seed depth was difficult to set when operating in summerfallow conditions because the softer soils allowed the seed depth to be affected by changes in the packing force and the depth of the cultivator.

Fertilizer placement of the knife opener was very good. The fertilizer was placed below and directly between the paired rows of seed in a row with a bandwidth of 0.5 in (13 mm). Depth of the fertilizer below the seed was variable.

Soil finishing of the packer attachment was good. Ridge depths left by the narrow spaced packers were an average of 3 in (76 mm) compared to 2 in (51 mm) ridge depths left by the wide spaced packers.

Trash clearance of the Morris Air Hoe Packer Attachment with the 5 in (127 mm) spacing was good and with the 3 in (76 mm) spacing was very good. The only plugging that occurred was in weed patches of fields or in stubble fields with large accumulations of trash.

Operation of the Morris Air Hoe Packer Attachment in stony conditions was very good. No damage occurred to the attachments from operation in stony conditions.

Ease of Operation and Adjustment: Ease of transporting the test cultivator with the Morris Air Hoe Packer Attachments mounted was fair. The cultivator had to be raised up to allow for more ground clearance.

Ease of setting the seed depth was fair. Seed depth was controlled by the vertical distance between the packer and the hoe opener. Ease of setting the packing force was poor. Packing force was controlled by the spring length or tension. Changing the tension required a small prybar to compress the spring.

The fertilizer boots did not plug during the test. The seed deflectors with the 3 in (76 mm) spaced packers would plug with soil when their packers stopped turning in very moist conditions. This was caused by a lack of clearance between the packer and the packer arm.

Power Requirements: Draft (drawbar pull) requirements depended on depth, field preparation, ground speed, soil type and moisture content. The Morris Air Hoe Packer Attachment pulled an average of 110% heavier than a 16 in (406 mm) sweep at the same seeding depth. This was with the fertilizer set 2 in (51 mm) deeper than the seed depth on the attachment. Maximum tractor power requirements in primary conditions for one air hoe packer attachment ranged from 4.8 hp to 8.6 hp (3.6 to 6.5 kW).

Ease of Installation: Ease of installing the Morris Air Hoe Packer Attachment was fair. Access to the nuts of the top U-bolts were obstructed by the springs. The entire test cultivator had to be raised up to accommodate the air hoe packer attachments.

Operator Safety: The Morris Air Hoe Packer Attachment presented two safety hazards during testing. Caution had to be taken when adjusting the spring tension on the packers and hitching the test cultivator was difficult due to the negative hitch weight on the cultivator.

Operator's Manual: The operator's manual was rated as good. It contained useful information on set up, operation and maintenance.

Mechanical History: Three of the wide sweeps broke during spring seeding. The top jam nuts on the seed adjustment bolts loosened during field operation. The knife openers were worn out at the end of the test.

Tractor Performance Monitors

Quality of Work: The Dickey-john TPM II tractor performance monitor was very useful in providing the operator with information on tractor operation.

Accuracy of the ground speed measurements was very good. Error was less than 1% if the monitor was properly calibrated for each field condition.

Accuracy of the wheel slip measurement was very good. Wheel slip was accurately displayed in 1% increments.

Accuracy of the area and workrate calculations was also very good. However, these values continued to increase in weedy fields on windy days, even though the tractor was stationary.

Ease of Installation: Ease of installation was very good. It took one man about four hours to install the Dickey-john TPM II. Installation instructions were clear and descriptive. The sensors were easily installed on the test tractor.

Ease of Operation and Adjustment: Ease of operating the Dickey-john TPM II was good after some operator experience had been gained.

Readability of the liquid crystal display was very good, but was reduced by glare from direct sunlight.

Ease of calibrating the monitor was good, however, the procedure was time-consuming since many calibration steps were required.

Electrical Power Requirements: The monitor drew a maximum 0.5 A when operating. Memory loss occurred if the tractor battery was disconnected or voltage dropped below 4 V.

Safety: Safety decals were provided to warn the operator against looking directly into the radar ground speed sensor.

Operator's Manual: The operator's manual was very good and was well illustrated.

Mechanical History: No mechanical problems occurred during the test.

Potato Planters

Summary of McConnell Model 555 Potato Planter

Overall functional performance of the McConnell Model 555 potato planter was very good.

The seed metering system was very good, which resulted in few misses or doubles.

Seed spacing was good. When set at a nominal spacing of 460 mm (18 in), 82% of the seed was spaced between 180 and 760 mm (7 and 30 in) with an average spacing of 465 mm (18 in). The coefficient of variation (CV) of single seed spacing was 31% at a speed of 8.6 km/h (5.3 mph). Acceptable seed spacing occurred at all speeds between 3 and 10 km/h (2 and 6 mph). Control of seeding depth was very good. Seed pieces were placed within 12 mm (0.5 in) of the desired depth. The closing disks were adjustable to produce very good hills with a uniform soil cover in all the soil conditions encountered.

Performance of the fertilizer metering system was good. Performance was reduced by the lack of available adjustment for low fertilizing rates. The application rate was not affected by field roughness, field slope, ground speed or fertilizer level in the hoppers. Fertilizer placement was good, with bands placed 50 mm (2 in) to each side of the seed. Opening disk adjustment was sufficient to permit the fertilizer bands to be placed below the seed level in all the soils encountered. No fertilizer calibration chart was supplied which necessitated calibrating the system.

Both the fertilizer and seed systems were convenient to adjust, service and operate. An adequate walkway was supplied for filling of the fertilizer hoppers but the seed hopper walkway was inadequate. No access holes were provided in the seed hoppers to enable the planter operator to break seed bridging.

A 65 kW (87 hp) tractor was adequate to operate the McConnell 555 in most soils at speeds up to 12 km/h (8 mph).

The operator's manual was good, and contained instructions on operation, adjustment, maintenance and safety. The McConnell was safe to operate if normal safety procedures were followed.

No major durability problems occurred during the test, although several minor problems occurred with the picker arms, hydraulic raising linkage and row markers.

Summary of Lockwood Model L06200-00423 Potato Planter

Overall functional performance of the Lockwood Model L06200-00423 potato planter was very good.

The seed feeding and metering system was very good, which resulted in few misses or doubles. Seed spacing was good. When set at a nominal spacing of 460 mm (18 in), 70% of the seed was spaced between 240 and 780 mm (9 and 30 in) with an average spacing of 490 mm (19 in). The coefficient of variation (CV) of single seed spacing was 28% at a speed of 9.4 km/h (6 mph). Acceptable seed spacing occurred at all speeds below 12 km/h (7 mph). Control of seeding depth was very good with seed pieces placed within 12 mm (0.5 in) of the desired depth. The closing disks could be adjusted to produce very good hills with uniform soil cover in all soil conditions encountered.

Performance of the fertilizer metering system was very good. The application rate was not affected by field roughness, field slope, ground speed or fertilizer level in the hoppers. Fertilizer placement was good, with bands placed 50 mm (2 in.) to each side of the seed. Opener adjustment was sufficient to permit the fertilizer bands to be placed below seed level in all soil conditions encountered. The fertilizer calibration chart was inaccurate necessitating recalibration for fertilizer commonly used in the prairie provinces.

Both the fertilizer and seed systems were convenient to adjust and service and were easy to operate. Adequate walkways were supplied for filling the seed and fertilizer hoppers. Access holes, provided in the back of the seed hoppers, were inadequate to enable the planter operator to conveniently break seed bridging.

A 65 kW (87 hp) tractor had adequate power to operate the Lockwood in most soils at speeds up to 10 km/h (6.2 mph).

The operator's manual was very good, containing comprehensive instructions on operation, adjustment, maintenance and safety. The Lockwood was safe to operate if normal safety procedures were followed.

No significant mechanical problems occurred during the test.

Summary of Dahlman Model PT Potato Planter

Overall functional performance of the Dahlman Model PT potato planter was very good.

The seed metering and feeding system was excellent which resulted in trouble free seed flow and few misses or doubles. Seed spacing was good. When set at a nominal spacing of 460 mm (18 in), 76% of the seed was spaced between 160 and 840 mm (6 and 33 in) with an average spacing of 440 mm (17 in). The coefficient of variation (CV) of single seed spacing was 38%. This coefficient was constant for all planting speeds. Control of seeding depth was very good with seed pieces placed within 12 mm (0.5 in) of the desired depth. The closing disks could be adjusted to produce very good hills with uniform cover in all soil conditions.

Performance of the fertilizer metering system was good. Performance was reduced by occasional feeding problems and exposure of the feed chains to rain. The application rate was not affected by field roughness, field slope, ground speed or fertilizer level in the hoppers. Fertilizer placement was good, with bands placed 130 mm (5 in) to each side of the seed. Opener adjustment was adequate to permit the fertilizer to be placed below seed level in all soil conditions. The manufacturer's fertilizer calibration chart was inaccurate which made it necessary to recalibrate the applicator.

Both the fertilizer and seed systems were convenient to adjust and service and easy to operate. An adequate walkway was supplied for filling the seed hoppers but the fertilizer hopper had no walkway.

A 60 kW (80 hp) tractor had adequate power to operate the Dahlman, in most soils, at speeds up to 10 km/h (6 mph).

The operator's manual was fair, containing instructions on operation, adjustment, maintenance and safety. The Dahlman was safe to operate if normal safety procedures were followed.

Several minor durability problems occurred with the picker arms and one castor wheel spindle had to be replaced during the test.

Summary of Acme 400 ST Potato Planter

Overall functional performance of the Acme 400 ST Potato Planter was very good.

The seed metering and feeding system was excellent, which resulted in trouble free seed flow and few misses or doubles. Seed spacing was good. When set at a nominal spacing of 460 mm (18 in), 72% of the cut seed was spaced between 220 and 860 mm (9 and 34 in), with an average spacing of 450 mm (18 in). The coefficient of variation (CV) of single seed spacing was 36%. This coefficient was constant for all planting speeds. The overall CV was 64% for cut seed and 66% for whole seed. Control of seeding depth was very good, with seed pieces placed within 12 mm (0.5 in) of the desired depth. The covering discs could be adjusted to produce excellent hills with uniform cover in all soil conditions.

Performance of the fertilizer metering system was good. Performance was reduced by occasional plugging of the flexible drop hoses and exposure of the feed screws to rain. The application rate was not affected by field roughness, field slope, ground speed or fertilizer level in the hoppers. Fertilizer placement was good, with bands placed 130 mm (5 in) to each side of the seed. Opener adjustment was adequate to permit the fertilizer to be placed below seed level in all soil conditions. The manufacturer's fertilizer charts were incomplete, making it necessary to recalibrate, to suit prairie application rates.

Both the fertilizer and seed systems were convenient to adjust, service and operate. No walkways were provided for filling of the seed or fertilizer hoppers.

A 100 kW (135 hp) tractor had adequate power to operate the Acme 400 ST in most soils, at speeds up to 10 km/h (6 mph).

The operator's manual was good, containing instructions on operation, adjustment, maintenance and safety. The Acme was safe to operate if normal safety procedures were followed.

Durability problems occurred with one castor wheel, and one cup chain jammed in the seed tube after jumping off a sprocket.

Summary of Lockwood L06200-00403 Accumatic Potato Planter

Overall functional performance of the Lockwood 403 Accumatic Potato Planter was very good. Performance in wet fields ranged from good to very good.

Performance of the electronic seed flow control system was very good, resulting in trouble-free seed flow to the picker bowls. Both seed hoppers emptied evenly. Seed spacing was good. When set at a nominal spacing of 460 mm (18 in), 54% of the seed was spaced between 230 and 690 mm (9 and 27 in), with an average spacing of 475 mm (19 in). The coefficient of variation (CV) of single seed spacing was 26%. This coefficient was fairly constant for all planting speeds. The overall CV was 64% when planting at 10 km/h (6 mph). Control of seeding depth was very good. The covering discs produced fair hills in most soil conditions.

The planting mechanism was very convenient to adjust, service and operate. No access was provided to enter the seed hopper for cleaning.

A 60 kW (80 hp) tractor would have adequate power to operate the Lockwood 403, in most soils, at speeds up to 10 km/h (6 mph), but a tractor with at least 100 kW (130 hp) is recommended. This size of tractor would provide better maneuverability and adequate floatation to carry the large hitch weight.

The operator's manual was very good, containing instructions on operation, adjustment, maintenance and safety. The Lockwood was safe to operate if normal safety procedures were followed.

The Lockwood transported well at speeds up to 25 km/h (15 mph) although transport ground clearance of 70 mm (3 in) was inadequate.

Operator visibility of the two outer pick wheels was very good, while visibility to the rear was poor. Maneuverability was very good, but use of tractor wheel brakes was often required on headlands.

No significant mechanical problems occurred during the test.

6

Education in Agricultural Engineering, Technology & Communication: Select Case Studies

Agricultural Education

Agricultural Education is an active part of the curriculum for many high schools in Indiana. This program area combines the home, the school, and the community as the means of education in agriculture. The courses provide students with a solid foundation of academic knowledge and ample opportunities to apply this knowledge through classroom activities, laboratory experiments and project applications, supervised agricultural experiences, and the FFA.

The vision and mission of Agricultural Education is: that all people value and understand the vital role of agriculture, food, fiber, and natural resource systems in advancing personal and global well-being; and that students are prepared for successful careers and a lifetime of informed choices in agriculture.

The goals for Agricultural Science and Business students focus on providing learning experiences that will allow them to:

- Demonstrate desirable work ethics and work habits.
- Apply the basic agricultural competencies and the basic background knowledge in agriculture and related occupations.
- Analyze entrepreneurial, business, and management skills needed by students preparing to enter agriculture and related occupations.
- Expand leadership and participatory skills necessary for the development of productive and contributing citizens in our democratic society.
- Gain effective social and interpersonal communication skills. Be aware of

career opportunities in agriculture and set career objectives.

- Acquire job-seeking, employability, and job-retention skills.
- Advance in a career through a program of continuing education and life-long learning.
- Apply the basic learning skills in reading, writing, thinking, mathematics, communicating, listening, and studying.
- Recognize the interaction of agriculture with governments and economic systems at the local, state, national, and international levels.
- Recognize how new technologies impact agriculture and how agriculture impacts the environment.

It is important to understand and reaffirm that career-technical experiences do not preclude students from going on to higher education; in fact participation actually enhances the opportunity. A growing number of students are combining both college preparation and work-place experiences in their high school preparation. Agricultural Science and Business and the FFA programs have a long history of successfully preparing students for entry level careers and further education and training in the science, business and technology of agriculture. The programs combine classroom instruction and hands-on career focused learning to develop students' potential for premier leadership, personal growth, and career success.

FFA

The FFA is the career and technical education student organization that is an integral part of the instruction and operation of a total agricultural education program. As an intra-curricular organization and essential component of the total program, the local agricultural education teacher(s) serve as the FFA chapter advisors. The many activities of the FFA parallel the methodology of the instructional program and are directly related to the occupational goals and objectives. As an integral part of the instructional program, district and state level FFA activities provide students opportunities to demonstrate their proficiency in the knowledge, skills, and attitudes they have acquired through the agricultural science and agricultural business total program. Agricultural education students demonstrating a high degree of competence in state level FFA activities are highly encouraged to represent their local communities, districts, and state by participating in national FFA activities.

Instructional activities of the FFA require participation of the agricultural science and agriculture business education students as an integral part of an agricultural education course of instruction and, therefore, may be considered an appropriate use and amount of the allotted instructional time.

Indiana Young Farmers' Association (IYFA)

The Indiana Young Farmers' Association was founded in 1962 and is dedicated to furthering the educational, social, and personal opportunities of all individuals interested or involved in the agricultural industry. It acts as an avenue for continuous educational experiences for adults, so that they can take full advantage of the possibilities available in the world of agriculture. IYFA supports the needs of agricultural-based, rural communities by providing worthwhile community services, effective leadership training, and wholesome social and recreational activities for the entire family through involvement in various local, state and national activities. It is the mission of the Young Farmer program to provide the opportunity to agriculturists to enter the industry as well as upgrade the skills needed to be leaders in their chosen occupation. Agriculture has become one of the most intense and technologically advanced industries in the world.

Exploring Agricultural Science And Business

The Agricultural Science and Business curriculum for middle level students follows the state standards of the Fundamentals of Agricultural Science and Business course. There is flexibility in content due to the length of the course offered locally. The primary objective is to introduce students to the dynamic industry of agriculture while gaining an awareness of the importance, impact, and diversity of careers in agricultural science and business. The content provides a hands-on exploratory, science-based approach to agriscience as well as providing a broad-based coverage of horticulture, animal science, environmental science, biotechnology, agricultural economics, plant and soil science, and agricultural science and agribusiness tools and equipment.

Fundamentals of Agricultural Science and Business

Fundamentals of Agricultural Science and Business is a yearlong course that is highly recommended as a prerequisite and foundation for all other agricultural classes. The nature of this course is to provide students with an introduction to careers and the fundamentals of agricultural science and business. Areas to be covered include: agricultural literacy, its importance and career opportunities, plant and soil science, environmental science, horticulture and landscape management, agricultural biotechnology, agricultural science and business tools and equipment, basic principles of and employability in the agricultural/horticultural industry, basic agribusiness principles and skills, developing leadership skills in agriculture, and supervised experience in agriculture/horticulture purposes and procedures. Student learning objectives are defined. Instruction includes not only agriculture education standards but many academic standards are included through the use of "hands-on" problem-solving individual and team activities.

Suggested Grade Levels: 9 or by permission of teacher. Recommended Prerequisite: None A two-credit/two semester course. May be offered as year long course to 8th

graders for high school credit. A Core 40, Core 40 with Academic Honors and Core 40 with Technical Honors diploma elective and directed elective course. A Career Academic Sequence, Career-Technical program, or Flex Credit course. Standards and learning activities defined. This course is included as a component of the Agriculture, Food and Natural Resources career cluster and may also be included as a component of the Building & Construction; Business, Management & Finance; Arts, A/V Technology & Communications; Health Services; and Science, Engineering & Information Technology career clusters.

Animal Science

This course is a yearlong program that provides students with an overview of the field of animal science. Students participate in a large variety of activities and laboratory work including real and simulated animal science experiences and projects. Areas that the students study may be applied to both large and small animals. Topics to be addressed include: anatomy and physiology, genetics, reproduction and biotechnology, nutrition, aquaculture, careers in animal science, animal health, meeting environmental requirements of animals, and management practices for the care and maintenance of animals.

- Suggested Grade Levels: 10-12
- Recommended Prerequisite: Fundamentals of Agricultural Science and Business or by permission of teacher
- A two-credit/ two-semester course. This course can be offered for a second full year at an advanced level and may also be offered in a two or three hour block with a maximum of six credit hours.
- This course may fulfill up to two credits of a life science requirement for the General Diploma.
- Animal Science may be offered as a small animal/large animal course and or include an advanced, local content specific application such as aquaculture.
- A Core 40, Core 40 with Academic Honors and Core 40 with Technical Honors diploma elective and directed elective course.
- A Career Academic Sequence, Career-Technical program, or Flex Credit course.
- Standards and learning activities defined.
- This course is included as a component of the Agriculture, Food and Natural Resources career cluster and may also be included as a component of the Science, Engineering & Information Technology; Personal & Commercial Services; Business, Management & Finance; Marketing, Sales & Promotion; and Health Services career clusters.

Food Science

This course is a yearlong program that provides students with an overview of food science and it importance. Introduction to principles of food processing, food chemistry and physics, nutrition, food microbiology, preservation, packaging and labeling, food commodities, food regulations, issues and careers in the food science industry help students understand the role that food science plays in the securing of a safe, nutritious, and adequate food supply. A project-based approach is utilized along with laboratory, team building, and problem solving activities to enhance student learning.

- Suggested Grade Levels: 11-12
- Recommended Prerequisite: Fundamentals of Agricultural Science and Business or by permission of the teacher
- A two-credit/two-semester course.
- This course may fulfill up to two credits of a life science requirement for the General Diploma.
- A Core 40, Core 40 with Academic Honors and Core 40 with Technical Honors diploma elective and directed elective course.
- A Career Academic Sequence, Career-Technical program, or Flex Credit course.
- Standards and learning activities defined.
- This course is included as a component of the Agriculture, Food and Natural Resources career cluster and may also be included as a component of the Business, Management & Finance; Health Services; Law, Public Safety, & Security; Marketing, Sales & Promotion; Science, Engineering & Information Technology; and Personal & Commercial Services career clusters.

Horticultural Science

Horticultural Science is a yearlong course designed to give students a background in the field of horticulture and its many career opportunities. It addresses the biology and technology involved in the production, processing, and marketing of horticultural plants and products. Topics covered include: reproduction and propagation of plants, plant growth, growth media, hydroponics, floriculture and floral design, management practices for field and greenhouse production, interior plantscapes, marketing concepts, production of herbaceous, woody, and nursery stock, fruit, nut, and vegetable production, integrated pest management and employability skills. Students participate in a variety of activities including extensive laboratory work usually in a school greenhouse.

- Suggested Grade Levels: 10-12
- Recommended Prerequisite: Fundamentals of Agricultural Science and Business or by permission of the teacher

- A two-credit/two-semester course. This course can be offered for a second full year at an advanced level and may also be offered in a two or three hour block for four semesters with a maximum of twelve credit hours.
- This course may fulfill up to two credits of a life science requirement for the General Diploma.
- A Core 40, Core 40 with Academic Honors and Core 40 with Technical Honors diploma elective and directed elective course.
- A Career Academic Sequence, Career-Technical program, or Flex Credit course.
- Standards and learning activities defined.
- This course is included as a component of the Agriculture, Food and Natural Resources career cluster and may also be included as a component of the Business, Management & Finance; Marketing, Sales & Promotion; and Transportation, Distribution & Logistics; Arts, A/V Technology & Communications; Science, Engineering & Information Technology; Social & Recreation Services career clusters.

Plant and Soil Science

Plant and Soil Science is a yearlong course that provides students with opportunities to participate in a variety of activities including laboratory work. Topics covered include: the taxonomy of plants, the various plant components and their functions, plant growth, plant reproduction and propagation, photosynthesis and respiration, environmental factors affecting plant growth, integrated pest management plants and their management, biotechnology, the basic components and types of soil, calculation of fertilizer application rates and procedures for application, soil tillage and conservation, irrigation and drainage, land measurement, grain and forage quality, cropping systems, precision agriculture, principles and benefits of global positioning systems and new technologies, harvesting, and career opportunities in the field of plant and soil science.

- Suggested Grade Levels: 10-12
- Recommended Prerequisite: Fundamentals of Agricultural Science and Business or by permission of the teacher
- A two-credit/two-semester course.
- This course may fulfill up to two credits of a physical, life, or earth/space science requirement for the General Diploma.
- A Core 40, Core 40 with Academic Honors and Core 40 with Technical Honors diploma elective and directed elective course.
- A Career Academic Sequence, Career-Technical program, or Flex Credit course.

- Standards and learning activities defined.
- This course is included as a component of the Agriculture, Food and Natural Resources career cluster and may also be included as a component of the Business, Management & Finance; Marketing, Sales & Promotion; and Transportation, Distribution & Logistics career clusters.

Agribusiness Management

Agribusiness Management is a yearlong course that presents the concepts necessary for managing an agriculture-related business from a local and global perspective. Concepts covered in the course include: exploring careers in agribusiness, global visioning, applying E-commerce, risk management, understanding business management and structures, entrepreneurship, the planning, organizing, financing, and operation of an agribusiness, economic principles, credit, computerized record keeping, budgeting, fundamentals of cash flow, federal, state, property and sales tax, insurance, cooperatives, purchasing, the utilization of information technology in agribusiness, marketing agricultural products, developing a marketing plan, advertising and selling products and services, understanding consumers and buying trends, agricultural law applications and employability skills.

- Suggested Grade Levels: 11-12
- Recommended Prerequisite: Fundamentals of Agricultural Science and Business or by permission of the teacher
- A two-credit/two-semester course.
- A Core 40, Core 40 with Academic Honors and Core 40 with Technical Honors diploma elective and directed elective course.
- A Career Academic Sequence, Career-Technical program, or Flex Credit course.
- Standards and learning activities defined.
- This course is included as a component of the Agriculture, Food and Natural Resources career cluster and may also be included as a component of the Business, Management, and Finance; Manufacturing & Processing; Marketing, Sales & Promotion; Personal & Commercial Services; Science, Engineering & Information Technology; Social & Recreation Services; and Transportation, Distribution & Logistics career clusters.

Agricultural Mechanization

Agricultural Mechanization is a yearlong, lab intensive course in which students develop an understanding of basic principles of selection, operation, maintenance, and management of agricultural equipment in concert with utilization of safety and technology.

Topics covered include: small and large gas and diesel engine repair, power transfer systems including hydraulic, pneumatic and robotic systems, arc, metal fabrication such as MIG, TIG and SMAW welding, concrete, wood, metal, electricity and electronics, recirculating aquaculture systems, hydroponics systems, surveying, precision farming equipment, remote sensing technology and global positioning systems equipment, building agriculture related buildings and structures including greenhouses, tillage, planting, irrigation, spraying, grain and forage harvesting, feed and animal waste management systems, agricultural industry communications and customer relations, safety and safety resources, career opportunities in the area of agricultural mechanization and employability skills.

- Suggested Grade Levels: 10-12
- Recommended Prerequisite: Fundamentals of Agricultural Science and Business or by permission of the teacher
- A two-credit/two-semester course. This course can be offered for a second full year at an advanced level and may also be offered in a two or three hour block for four semesters with a maximum of twelve credit hours.
- A Core 40, Core 40 with Academic Honors and Core 40 with Technical Honors diploma elective and directed elective course.
- A Career Academic Sequence, Career-Technical program, or Flex Credit course.
- Standards and learning activities defined.
- This course is included as a component of the Agriculture, Food and Natural Resources career cluster and may also be included as a component of the Building & Construction; Manufacturing & Processing; Mechanical Repair & Precision Crafts; and Science, Engineering & Information Technology career clusters.

Farm Management

Farm Management is a yearlong course that introduces students to the principles of farm organization and management with the utilization of technology. It covers the effects of good and poor management on a farm, economic principles, decision-making, methods for organizing and planning, getting started in the farming business, farm record keeping systems, risk management, and career opportunities in the field of farm management.

- Suggested Grade Levels: 11-12
- Recommended Prerequisite: Fundamentals of Agricultural Science and Business or by permission of the teacher

- A two-credit/two-semester course.
- A Core 40, Core 40 with Academic Honors and Core 40 with Technical Honors diploma elective and directed elective course.
- A Career Academic Sequence, Career-Technical program, or Flex Credit course.
- Standards and learning activities defined.
- This course is included as a component of the Agriculture, Food and Natural Resources career cluster and may also be included as a component of the Business, Management, and Finance; Marketing, Sales & Promotion career clusters.

Landscape Management

Landscape Management is a yearlong course that provides the student with an overview of the many career opportunities in the diverse field of landscape management. Students are introduced to the procedures used in the planning and design of a landscape using current technology practices, the principles and procedures involved with landscape construction, the determination of maintenance schedules, communications, management and employability skills necessary in landscaping operations, and the care and use of equipment utilized by landscapers. Upon completion of the program, students have the opportunity to receive an industry approved State Certificate of Mastery in Landscape Management.

- Suggested Grade Levels: 10-12
- Recommended Prerequisite: None
- A two-credit/two-semester course. This course can be offered for a second full year at an advanced level and may also be offered in a two or three hour block for four semesters with a maximum of twelve credit hours.
- A Core 40, Core 40 with Academic Honors and Core 40 with Technical Honors diploma elective and directed elective course.
- A Career Academic Sequence, Career-Technical program, or Flex Credit course.
- Standards and learning activities defined.
- This course is included as a component of the Agriculture, Food and Natural Resources career cluster and may also be included as a component of the Arts, A/V Technology & Communications; Building & Construction; Marketing, Sales & Promotion; and Business, Management and Finance career clusters.

Natural Resource Management

This course is a yearlong program that provides students with a background in natural resource management. Students are introduced to career opportunities in natural resource management and related industries, understanding forest ecology importance, recognizing trees and their products, tree growth and development, forest management, measuring trees, timber stand improvement and urban forestry, soil features, erosion and management practices, conservation practices, water cycles, uses, quality standards, reducing water pollution, conducting water quality tests, watersheds, and its importance to natural resource management, hazardous waste management, native wildlife, waterfowl, wetlands, and fish management, topography map use, management of recreational areas, game bird and animal management, outdoor safety, and weather. "Hands-on" learning activities encourage students to investigate areas of environmental concern including: identification and management of ecosystems, natural succession identification, natural communities, recycling and management of waste in the environment, soil conservation management practices, land uses, and air quality.

- Suggested Grade Levels: 10-12
- Recommended Prerequisite: None
- A Core 40, Core 40 with Academic Honors and Core 40 with Technical Honors diploma elective and directed elective course.
- A Career Academic Sequence, Career-Technical program, or Flex Credit course.
- Standards and learning activities defined.
- This course is included as a component of the Agriculture, Food and Natural Resources career cluster and may also be included as a component of the Arts, A/V Technology & Communications; Business, Management & Finance; Law, Public Safety & Security; Manufacturing & Processing; Marketing, Sales & Promotion; Social & Recreation Services; Transportation, Distribution & Logistics career clusters.

Supervised Agricultural Experience (SAE)

CIP Code: SAE Cooperative Education –Funded By -Ag CIP Appropriate for training site

CIP Code: Non-Cooperative – Not funded by state but may be offered for credit

Supervised Agricultural Experience (SAE) is designed to provide students with opportunities to gain experience in the agriculture field(s) in which they are interested. Students should experience and apply what is learned in the classroom, laboratory, and training site to real-life situations. Students work closely with their agricultural science and business teacher(s), parents, and/or employers to get the most out of their SAE program. This course can be offered each year as well as during the summer session. SAE may be offered as a Cooperative Education Program. Curriculum content and competencies should be varied so that school year and summer session experiences are not duplicated.

- Suggested Grade Levels: 10-12
- Recommended Prerequisite: Fundamentals of Agricultural Science and Business
- A maximum of eight credits can be earned in this course when offered as a one hour course/eight semesters, some of which can be earned during summer sessions. Curriculum content and competencies should not be duplicated when multiple credits are being earned.
- As Cooperative Education: one credit related instruction and two credits on the job training/four semesters = 12 credit hours (on the job training credit hours may be increased in approved situations).
- A Core 40, Core 40 with Academic Honors and Core 40 with Technical Honors diploma elective and directed elective course.
- A Career Academic Sequence, Career-Technical program, or Flex Credit course.
- Standards and learning activities defined.
- This course is included as a component of the Agriculture, Food and Natural Resources career cluster and may also be included as a component of the Arts, A/V Technology & Communications; Building & Construction; Education & Training; Law, Public Safety & Security; Manufacturing & Processing; Marketing, Sales & Promotion; Mechanical Repair & Precision Crafts; Personal & Commercial Services; Science, Engineering & Information Technology; Social & Recreation Services; Transportation, Distribution & Logistics; Business, Management and Finances; and Health Services career clusters.

Advanced Life Science, Animals (L)

Advanced Life Science, Animals, is a standards-based, interdisciplinary science course that integrates biology, chemistry, and microbiology in an agricultural context. Students enrolled in this course formulate, design, and carry out animal-based laboratory and field investigations as an essential course component. Students investigate key concepts that enable them to understand animal growth, development and physiology as it pertains to agricultural science. This course stresses the unifying themes of both biology and chemistry as students work with concepts associated with animal taxonomy, life at the cellular level, organ systems, genetics, evolution, ecology, and historical and current issues in animal agriculture. Students completing this course will be able to apply the principles of scientific inquiry to solve problems related to biology and chemistry in highly advanced agricultural applications of animal development.

- Suggested Grade Levels: 11-12
- Highly Recommended Prerequisite: Biology and Chemistry due to course content standards
- A two semester course, one credit per semester
- Counts as two credits of a Core 40 life science course.
- A Core 40, Core 40 with Academic Honors and Core 40 with Technical Honors diploma elective and directed elective course.
- A Career Academic Sequence, Career-Technical program, or Flex Credit course.
- Content standards and competencies are defined.
- Standards and learning activities defined.
- This course is included as a component of the Agriculture, Food and Natural Resources career cluster and may also be included as a component of the Science, Engineering & Information Technology; Personal & Commercial Services; Business, Management & Finance; Marketing, Sales & Promotion; and Health Services career clusters.

Advanced Life Science, Plant and Soil (L)

Advanced Life Science, Plant and Soil, is a standards-based, interdisciplinary science course that integrates the study of advanced biology, chemistry, and earth science in an agricultural context. Students enrolled in this course formulate, design, and implement agriculturally-based laboratory and field investigations as an essential course component. These extended laboratory and literature investigations focus on the chemical reactions of matter in living and nonliving materials while stressing the unifying themes of chemistry and the development of physical and mathematical models of matter and its interactions. Using the principles of scientific inquiry, students examine the internal structures, functions, genetics and processes of living plant organisms and their interaction with the environmental. Students completing this course will be able to apply the principles of scientific inquiry to solve problems related to both biology and chemistry in the context of highly advanced agricultural applications of plants and soils.

- Highly Recommended Prerequisite: Biology and Chemistry due to course content standards
- A two semester course, one credit per semester
- Counts as two credits of a Core 40 life science course.

- A Core 40, Core 40 with Academic Honors and Core 40 with Technical Honors diploma elective and directed elective course.
- A Career Academic Sequence, Career-Technical program, or Flex Credit course.
- Content standards and competencies are defined.
- This course is included as a component of the Agriculture, Food and Natural Resources career cluster and may also be included as a component of the Business, Management & Finance; Marketing, Sales & Promotion; and Transportation, Distribution & Logistics career clusters.

Advanced Life Science, Foods (L)

Advanced Life Science, Foods, is a standards-based, interdisciplinary science course that integrates biology, chemistry, and microbiology in an agricultural context. Students enrolled in this course formulate, design, and carry out food based laboratory and field investigations as an essential course component. Students understand how biology, chemistry, and physics principles apply to the composition of foods, food nutrition and development, food processing, and storage. Students completing this course will be able to apply the principles of scientific inquiry to solve problems related to biology, physics and chemistry the context of highly advanced agricultural applications of food.

- Highly Recommended Prerequisite: Biology and Chemistry due to course content standards
- A two semester course, one credit per semester
- Counts as two credits of a Core 40 life science course.
- A Core 40, Core 40 with Academic Honors and Core 40 with Technical Honors diploma elective and directed elective course.
- A Career Academic Sequence, Career-Technical program, or Flex Credit course.
- Content standards and competencies are defined.
- This course is included as a component of the Agriculture, Food and Natural Resources career cluster and may also be included as a component of the Business, Management & Finance; Health Services; Law, Public Safety, & Security; Marketing, Sales & Promotion; Science, Engineering & Information Technology; and Personal & Commercial Services career clusters.

Agricultural Education: A Case Study of Utah State University

The Bachelor of Science (BS) Degree in Agricultural Education is designed to develop the needed background knowledge of basic agricultural concepts and technologies for teaching and cooperative extension. Coursework covers topics in such fields as agricultural economics; agricultural mechanization; animal, dairy and veterinary sciences; plant and soil science; and natural resources. Students choosing this major must also be admitted to the Secondary Teacher Education Program (STEP) administered by the College of Education and Human Services.

Curriculum in Agricultural Engineering: A Case Study of Urbane

Agricultural engineering is the application of science, mathematics, and engineering to agriculture, food systems, natural resources, the environment, and related biological systems. This program has special emphasis on environmental protection and the biological interface of plants, animals, and soils with the design and performance of environments, machines, mechanisms, processes, and structures.

Educational Objectives

The agricultural engineering program develops graduates with a comprehensive education for engineering careers in industry, consulting, government, and academia. The program provides a combination of courses in engineering theory and application for students seeking careers in agricultural production, bioprocess systems, environmental protection, and product utilization.

Students will be provided the opportunity to:

- develop knowledge and appreciation of biological and agricultural systems as a basis for engineering applications,
- identify and analyze engineering problems related to these systems and organize a systematic approach to their solution and evaluation,
- design machines, systems, and processes considering economic, social, and environmental impacts,
- develop communication, learning, and teamwork skills,
- develop a sense of social and professional responsibility for solving engineering problems that benefit society.

Educational Outcomes

Engineering design, teamwork, and communication are integrated throughout the curriculum, culminating in a capstone design experience. By choice of electives, students not only meet all the outcomes listed in the College of Engineering introductory description but are also provided with the opportunity to specialize in bioenvironmental engineering, food and bioprocess engineering, grain quality and milling properties, off-road equipment engineering, precision agriculture, or soil and water resource engineering. The curriculum requires 128 hours for graduation, except for the specialization in food and bioprocess engineering, which requires 132 hours for graduation.

For more details or information regarding the curriculum, visit our Web site or arrange a personal visit to the department. Specialization in Bioenvironmental Engineering, Off-Road Equipment Engineering, or Soil and Water Resource Engineering

First year

Hours	*First Semester*
1	AG E 100-Introduction to Agricultural Engineering
3	CHEM 101-General Chemistry
1	CHEM 105-General Chemistry Laboratory
0	ENG 100-Engineering Lecture
3	G E 103-Engineering Graphics and Desig 5 MATH 120-Calculus and Analytic Geometry, I
4	RHET 105-Principles of Composition[1]
17	Total

Hours	*Second Semester*
3	CHEM 102-General Chemistry (Biological or Physical Version)*
1	CHEM 106-General Chemistry Laboratory (Biological or Physical Version)*
3	MATH 130-Calculus and Analytic Geometry, II
2	MATH 225-Introductory Matrix Theory
4	PHYCS 111-General Physics (Mechanics)
4	Biological and natural sciences elective[2]
17	Total

*Biological version recommended.

Second year

Hours	*First Semester*
4	AG E 221-Engineering for Agricultural and Biological Systems
3	C S 101-Introduction to Computing with Application to Engineering and Physical Science
3	MATH 242-Calculus of Several Variables
4	PHYCS 112-General Physics (Electricity and Magnetism)
2-3	T A M 150-Introduction to Statics or T A M 152-Engineering Mechanics, I-Statics
16-17	Total

Hours	Second Semester
4	AG E 222-Engineering for Bioprocessing and Bioenvironmental Systems
3	MATH 285-Differential Equations and Orthogonal Functions
2	PHYCS 113-General Physics (Fluids and Thermal Physics)
3	T A M 212-Engineering Mechanics, II-Dynamics
3	Elective in social sciences or humanities[3, 4]
15	Total

Third year

Hours	First Semester
3	Agricultural engineering technical elective[5]
3	ECE 205-Introduction to Electric and Electronic Circuits
1	ECE 206-Introduction to Electric and Electronic Circuits Laboratory
3	T A M 221-Introduction to Solid Mechanics
3-4	STAT 310/MATH 363-Introduction to Mathematical Statistics and Probability, I; or CEE 293-Engineering Modeling Under Uncertainty; or I E 230-Analysis of Data College of Engineering
3	Elective in social sciences or humanities[3, 4]
16-17	Total

Hours	Second Semester
3	Agricultural engineering technical elective[5]
1	AG E 298-Undergraduate Seminar
3	ECON 103-Macroeconomic Principles[3]
3-4	M E 205-Thermodynamics, or CH E 370-Chemical Engineering Thermodynamics
3-4	T A M 235-Introduction to Fluid Mechanics, or CH E 371-Fluid Mechanics and Heat Transfer, or M E 211-Introductory Gas Dynamics
3	Elective in social sciences or humanities[3, 4]
16-18	Total

Fourth year

Hours	*First Semester*
3	**Agricultural engineering technical elective[5]**
3	**Elective in social sciences or humanities[3, 4]**
4	**Technical elective[5]**
3	**Free elective[4]**
2	**AG E 299-Undergraduate Thesis**
15	**Total**

Hours	*Second Semester*
3	**Agricultural engineering technical elective[5]**
3	**Free elective[4]**
3	**Technical elective[5]**
4	**Biological and natural sciences elective[2]**
3	**Elective in social sciences or humanities[3, 4]**
16	**Total**

1. Students may take SPCOM 111 and 112 in place of RHET 105. RHET 105 may be taken in the first or second semester of the first year.

2. Students must complete eight hours from biological and natural sciences approved list.

3. Each student must satisfy the social sciences and humanities requirements of the College of Engineering, including ECON 102 or 103, and the campus general education requirements to social sciences and humanities.

4. One elective course must satisfy the Eeneral Education Advanced Composition requirement.

5. Students must have 19 hours of technical electives; at least 12 hours must be from AG E courses and the remainder selected from the department-approved list.

Biological and Natural Sciences Electives

Hours	
8	**Choose from (8 hours minimum):**
4	**AN SCI 202-Domestic Animal Physiology**
3	**AN SCI 307-Environmental Aspects of Animal Management**

3	BIOL 100-Biological Sciences[1]
4	BIOL 101-Biological Sciences[1]
4	BIOL 104-Animal Biology[1]
3	CHEM 231-Elementary Organic Chemistry, I
2	CHEM 234-Elementary Organic Chemistry Laboratory, I
3	CPSC/ENT/NRES 120-Introduction to Applied Entomology
3	CPSC 322-Forage Crops and Pastures
3	FSHN 371/MCBIO 311-Food and Industrial Microbiology
4	GEOL 101-Introduction to Physical Geology
3	GEOL 250-Geology for Engineers
3	MCBIO 100-Introductory Microbiology[1]
2	MCBIO 101-Introductory Experimental Microbiology
2	MCBIO 312-Techniques of Applied Microbiology
4	NRES 101-Introductory Soils
3	NRES 245-Indoor Plant Culture, Use, and Identificatio
4	NRES 365-Growth and Development of Horticultural Crops
4	PLBIO 100-Plant Biology[1] or CPSC 121-Principles of Field Crop Production and Protection
4	PHYSL 103-Introduction to Human Physiology

1. Students must take at least one of these courses.

Technical Electives

For a total of 19 hours.

Agricultural Engineering Technical Electives

Hours	
3	AG E 236-Machine Characteristics and Mechanisms
3	AG E 277-Design of Architectural Structures[1]
3	AG E 287-Environmental Control for Plants and Animals[1]
3-4	AG E 311-Instrumentation and Measurement[2]
3	AG E 315-Applied Machine Visio 3 AG E 336-Engineering Design Projects for Agricultural Industries[1]
3	AG E 346-Tractors and Prime Movers

3	AG E 356-Soil and Water Conservation Structures[1]
3	AG E 357-Land Drainage[1]
3	AG E 360-Indoor Air Contaminant Measurement and Control
3	AG E 383-Engineering Properties of Food Materials
2	AG E 385-Food and Process Engineering Design[1]
3	AG E 387-Grain Drying and Conditioning
3	AG E 389-Process Design for Corn Milling

1. Students must take at least one of these courses. Includes major design experience.
2. This course is strongly recommended.

Other Technical Electives

Choose the remainder of the 19 hours from:

4	CEE 201-Engineering Surveying or CE 205-Route Surveying and Desig
3	CEE 241-Environmental Quality Engineering
3	CEE 255-Introduction to Hydrosystems Engineering[1]
3	CEE 261-Introduction to Structural Engineering[1]
3	CEE 263-Behavior and Design of Metal Structures,I
3	CEE 264-Reinforced Concrete Design, I
3	CEE 280-Introduction to Soil Mechanics and Foundation Engineering
3	CEE 350-Surface Water Hydrology
3	CEE 361-Methods of Structural Analysis
3	CH E 261-Introduction to Chemical Engineering
4	CH E 370-Chemical Engineering Thermodynamics
4	CH E 371-Fluid Mechanics and Heat Transfer
4	CH E 373-Mass Transfer Operations
3	G E 288-Engineering Economy and Operations Research
4	M E 231-Engineering Materials
3	M E 271-Mechanical Design, I[1]
3	M E 285-Design for Manufacturability
3	M E 307-Solar Energy Υτιλιζατιον
3	MFG E 210-Introduction to Manufacturing Systems
3	MFG E 350-Information Management for Manufacturing Systems

5	PHYCS 343/CHEM 323-Electronic Circuits, I
	Any 200- or 300-level engineering course approved by an adviser.

1. One of these courses is strongly recommended.

Specialization in Food and Bioprocess Engineering

Food and bioprocess engineering is the application of engineering principles to produce, preserve, process, package, and distribute foods. Food and bioprocess engineers develop, design, and construct new machinery, processes, and plants; they develop and test new products; they preserve and distribute foods; and they manage environmental factors, waste products, and energy. Food and bioprocess engineers participate in nearly every phase of food processing. Graduates are prepared for positions in a variety of industries, including food, pharmaceutical, and biotechnology industries. Job opportunities also exist with the government, universities, and consulting firms. Career possibilities include research and development; project, process, and plant engineering, which can include design, optimization, and construction; technical sales and service; and supervision and management. Those who continue their education in graduate school will have a strong background for further study in the sciences or engineering.

First year

Hours	*First Semester*
1	AG E 100-Introduction to Agricultural Engineering
3	CHEM 101-General Chemistry
1	CHEM 105-General Chemistry Laboratory
0	ENG 100-Engineering Lecture
3	G E 103-Engineering Graphics and Desig
5	MATH 120-Calculus and Analytic Geometry, I
4	RHET 105-Principles of Composition[1]
17	Total

Hours	*Second Semester*
3	CHEM 102-General Chemistry (Biological or Physical Version)
1	CHEM 106-General Chemistry Laboratory (Biological or Physical Version)
3	C S 101-Introduction to Computing with Application to Engineering and Physical Science
3	MATH 130-Calculus and Analytic Geometry, II

2	MATH 225-Introductory Matrix Theory
4	PHYCS 111-General Physics (Mechanics)
16	Total

Second Year

Hours	*First Semester*
3	CHEM 231-Elementary Organic Chemistry, I
3	ECON 103-Macroeconomic Principles[2]
3	MATH 242-Calculus of Several Variables
3	MCBIO 100-Introductory Microbiology
2	PHYCS 113-General Physics (Fluids and Thermal Physics)
2-3	T A M 150-Introduction to Statics or T A M 152-Engineering Mechanics, I-Statics
16-17	Total

Hours	*Second Semester*
4	AG E 222-Engineering for Bioprocessing and Bioenvironmental Systems
3	MATH 285-Differential Equations and Orthogonal Functions
2	MCBIO 101-Introductory Experimental Microbiology
4	PHYCS 112-General Physics (Electricity and Magnetism)
3	T A M 212-Engineering Mechanics, II-Dynamics
16	Total

Third Year

Hours	*First Semester*
3	CH E 261-Introduction to Chemical Engineering
4	FSHN 314-Food Chemistry and Nutrition, I
3	T A M 221-Introduction to Solid Mechanics
6	Electives in social sciences or humanities[2, 4]
16	Total

Hours	*Second Semester*
1	AG E 298-Undergraduate Seminar
4	CH E 370-Chemical Engineering Thermodynamics

3	ECE 205-Introduction to Electric and Electronic Circuits
3	FSHN 371/MCBIO 311-Food and Industrial Microbiology
3	Free elective[4]
3	Elective in social sciences or humanities[2, 4]
17	Total

Fourth Year

Hours	*First Semester*
3	AG E 383-Engineering Properties of Food Materials
4	CH E 371-Fluid Mechanics and Heat Transfer
3	FSHN 361-Food Processing, I
4	Technical elective[3]
3	Elective in social sciences or humanities[2, 4]
17	Total

Hours	*Second Semester*
2	AG E 299-Undergraduate Thesis
2	AG E 385-Food and Process Engineering Desig 4 CH E 373-Mass Transfer Operations
3	FSHN 362-Food Processing, II
3	Free elective[4]
3	Elective in social sciences or humanities[2, 4]
17	Total

1. Students may take SPCOM 111 and 112 in place of RHET 105. RHET 105 may be taken in the first or second semester of the first year.

2. Each student must satisfy the social sciences and humanities requirements of the College of Engineering, including ECON 102 or 103, and the campus general education requirements to social sciences and humanities.

3. Students select technical electives from the approved list for food and bioprocess engineering.

4. One elective course must satisfy the General Education Advanced Composition requirement.

Food and Bioprocess Engineering Electives

Hours	*Technical Electives*
3-4	AG E 311-Instrumentation and Measurements
3	AG E 315-Applied Machine Visio
3	AG E 387-Grain Drying and Conditioning

3	AG E 389-Process Design for Corn Milling
3	AG E 396-Special Problems (Package Engineering)
3-4	CEE 293-Engineering Modeling Under Uncertainty, I E 230-Analysis of Data, or STAT 310/MATH 363-Introduction to Mathematical Statistics and Probability, I
3	CH E 389-Chemical Process Control and Dynamics
3	G E 288-Engineering Economy and Operations Research
3.5	M E 261-Fundamentals of Signal Processing, Instrumentation, and Control
3	M E 271-Mechanical Design, I
2	MCBIO 312-Techniques of Applied Microbiology
	Other electives, subject to approval.

Agricultural Machinery Technology at Utah State University: A Case Study

The Agricultural Machinery Technology Program, which includes both the One-year Technology Certificate and the Associate of Applied Science Degree (two-year) Program, is the only one of its kind in the Western United States. This critical scarcity has secured a program endorsement from the Implement Dealers' Association, which represents dealers and manufacturers throughout the Western United States.

One-year Technology Certificate Program. The One-year Technology Certificate Program has its basis in agricultural machinery. This program is designed to familiarize students with the agricultural equipment industry. The Machinery Certificate includes technical hands-on training on engines, power trains, hydraulics, DC electrical, and specialized forage, harvesting, tillage, planting, and spraying equipment. At the end of the year, students participate in an occupational experience with a cooperating employer. This employment may extend on through the summer, developing excellent opportunities for future employer reference and on-the-job experience.

Associate of Applied Science Degree. The Associate of Applied Science Degree is a second-year continuation of the one-year program in agricultural machinery technology. In this program, specialization courses and training are designed to allow maximum flexibility in developing a curriculum for the student. Basic core classes must be completed in equipment testing, diagnosis, and retailing of parts and equipment. Because it is an associate degree program, a minimum of 6 credits of University Studies courses are required. This requirement facilitates the transition into the four-year program for those students who decide to continue with their education.

Career Opportunities — national and international

- The One-year Technology Certificate Program provides preparation for entry-level employment opportunities with agricultural implement dealerships and companies in the areas of parts and service. Other opportunities include employment with farm suppliers, feed and fertilizer agencies, corporate farms and ranches, and other related enterprises in positions as salespersons, parts clerks, service center foremen, machinery fieldmen, and assemblymen. The department averages six employment inquiries from dealers and manufacturers for every student available for placement.
- The Associate of Applied Science Degree is designed to prepare students for rapid advancement in the fields of service management or parts management in positions as sales managers, parts managers, and service managers. These students may be employed by manufacturers such as AGCO, Cat, CNH, John Deere, and Kubota.

Agricultural Systems Technology at Utah State University: A Case Study

Agricultural Systems Technology (AST) prepares individuals to manage businesses with the application of sound technical, economical, and environmental practices. Leading-edge technology is applied to real-world problems, with the emphasis on being productive. Each program within the baccalaureate degree emphasizes technical information, applied research, and a systems approach to solving problems.

Problem solving practices rely on a basic strategy that includes problem identification, collection of information, and evaluating solutions to achieve a goal. The AST program tackles these concepts in a formal manner, using computer software, information technology, and mathematics. The AST major provides real-world instruction with resources that afford the greatest balance between leading-edge technology and its full, productive use.

Agricultural Systems Technology prepares individuals to organize and manage environmentally sound, technology-based businesses. The emphasis is on planning and directing industry or business projects, placing emphasis on results.

Career Opportunities — national and international

- manufacturing and processing operations
- advice and trouble-shooting help on technical equipment (or projects)
- planning buildings and equipment to fit and work together, working with the handling and flow of materials such as grain, feeds, chemicals, vegetables, fruits, etc. and products made from them

- applying technical training in selling or product demonstration
- teaching people about product use and value
- managing and operating farms or agri-businesses
- sustainable agriculture practices.

Agricultural Communication and Journalism at Utah State University: A Case study

Agricultural communication and journalism is a field of study designed to transfer technological ideas and change people's behaviors. The agricultural production and processing industry is faced with issues of image, ethics and survival. Production systems have increased capacities that enable farmers and ranchers to produce the commodities to feed this nation and much of the world's population. Processors, marketers and distributors have demonstrated a high level of efficiency. Yet, American agriculture comes under fire from consumers, government agencies, and environmental interest groups within this country and internationally. The greatest deficiency of agricultural production and processing is the effective communication of ideals and standards to wide and varied audiences. A significant demand exists in agribusiness and related fields for skilled communicators. This program combines courses in journalism and agriculture to develop a well-rounded agricultural communications professional.

If you don't have the Adobe Acrobat reader left click on the image below to go to the Adobe site for a free download.